RENEE DANG

HARVESTING RAINWATER

FOR YOUR HOMESTEAD
IN 9 DAYS OR LESS

7 Steps to Unlocking Your Family's Clean, Independent, and Off-Grid
Water Source with the QuickRain Blueprint

Editing by Lisa Howard

Illustrations by Liu Lu

Interior Layout Design by Dimitra Kostarelou

Moral Support and Bubble Tea by Jeffrey Ward

TABLE OF CONTENTS

A free gift to our readers

Learn the 9 Mistakes Contaminating Your Rainwater Right Now (and how to fix them!) Grab your free gift by visiting:

www.reneedang.com

 Or, scan here with your phone!

INTRODUCTION

H ave you suddenly realized that your homestead is having trouble with its water source? Are you finding that a well is extremely expensive to dig where you live or that you have no groundwater? Maybe you've just always wondered how you could collect and use the free rain falling from the sky as a resource for you and your family. This book will show you how to harvest rain and start using it for all of your needs quickly, easily, and without breaking the bank.

Harvesting rainwater offers many advantages. In recent years (especially with greater water shortages and tougher droughts), families and farmers have been turning back to the world's primary source of fresh water—rain—when they realized that their groundwater and rivers were drying up. Laws and legislation in the US against rainwater harvesting are quickly changing to incentivize rather than discourage harvesting rain. The City of Philadelphia will pay up to $100,000 per acre[1] of rainwater captured! Perhaps your local state or municipality is also accommodating the increasing demand for legal rainwater harvesting.

You may also be incentivized to collect rainwater due to limited or poor groundwater resources. Many families in the US want to have a backup source of water if their well runs dry. You may share a well with other families, but you may also be looking for greater control of your water source—while everyone can do their best to follow the rules and not contaminate a well, it just takes a few mistakes to spoil the well for an entire water table. But if you collect rain in tanks that only your family manages, you will once again find the peace of mind that comes with having an independent water source.

Rainwater is a distilled, nutrient-filled, and—depending on where you live—rather abundant resource. (It should be noted that many desert homesteaders in the southwestern United States collect rainwater year-round as their primary source of water. And they live in a place generally considered to be extremely arid! Does your location receive more rain than a desert?) If properly treated, rainwater can be used for drinking and cooking, and it can also be used for gardening. Rainwater itself is no more acidic than coffee, and its slight acidity and high oxygen and nitrogen content create very green, healthy, strong plants. Whether they're indoor or outdoor, you can help your plants fight disease, yield more fruit, and grow stronger root systems by watering them with rainwater. Watering livestock such as chickens or goats with rainwater is another great way to use rain.

This book will take a beginner with zero knowledge of rainwater harvesting to feeling confident in sourcing, installing, and maintaining the right rainwater harvesting system for their garden or homestead in 9 days or less. The primary focus of this book will be on using tank systems to collect rainwater for later use.

THE SIMPLE TRUTH ABOUT HARVESTING RAINWATER

Rainwater harvesting is simple: collect the rain, store the rain, and use the rain. Harvesting can be optimized by taking into account the topography of your location, how much space you have, which type of pump to get, which materials will optimize rain catchment efficiency…the list goes on. Having so many options can make harvesting seem like an endless and daunting task. However, most rainwater harvesters follow the same basic principles, plans, and system requirements to meet their needs. This book distills all of the considerations that pertain to safety, installation, and maintenance into a handful of guiding principles to get you collecting rain right away.

While your specific topography or climate may require slight changes in the size or shape of your system (as well as careful winterization or frost protection considerations for cold climates), the methods described in this book can be adapted by most rainwater harvesters. Collectively, these methods are called the QuickRain Blueprint.

The QuickRain Blueprint uses the most common, the most recommended, and the easiest-to-maintain type of system and scales it to meet your family's water needs. This is a "dry system," and it routes rain from the roof to the gutters and then straight into a tank. There's no need to bury pipes or tanks. It's the quickest system to set up and maintain, which is exactly what's most beneficial to those needing a reliable water source—fast.

A dry system basically works as an enlarged rain barrel. It works beautifully for most people, most homes, and most properties. It may work beautifully for you and your family as well. This method simplifies rainwater harvesting: there's no need to bury pipes, no need to worry about pipes bursting in the winter, and no need to worry about inspecting an underground tank. The QuickRain Blueprint uses the roof and gutters you already have, and it guides you through selecting, scaling, and sourcing your materials to installing and maintaining your system so that you can feel confident in your water quality. The QuickRain Blueprint can be adapted for chicken coops, sheds, or any other roofed structure on your property. The math is already done for you! You just need to choose the size of the tank(s) you need depending on your water needs, your budget, and how much space you have. Because of how quick and easy the QuickRain Blueprint is, the systems in this book can be chosen and installed in as quickly as 9 days or less (not accounting for variable delivery or fulfillment times for parts or materials).

The QuickRain Blueprint involves following 7 easy steps, each of which has its own chapter in the book:

1. Learn the legalities of your locality
2. Learn the 10 basic components of a rainwater harvesting system
3. Decide on the end-use goals for your rainwater
4. Learn the basics of rainwater harvesting safety and use
5. Scale and source your system
6. Install your system
7. Maintain your system

Before you can build your system, you must first learn the basics of rainwater harvesting systems, such as the rainwater harvesting regulations in your area and the components of every rainwater harvesting system. Armed with this knowledge, you'll be prepared to figure out the size of the system you should build, where to acquire the equipment and components you'll need for the system, how to install it, and how to maintain it. By doing all of this, you'll gain a foundation of knowledge that you can use to confidently install a rainwater harvesting system of any size on your property. Learn the basics, and the world of rainwater harvesting will be open to you!

CHAPTER 1

LEARN THE LEGALITIES
OF YOUR LOCALITY |

T he first step on your rainwater harvesting journey, especially if you live in the US, is to learn the specific laws related to rainwater harvesting in your state and local area, whether that be your county or municipality. It is not illegal to harvest rain in any state—while there are certainly constraints you need to abide by, you may have more leeway than you think. Researching what the exact regulations are will help clear up any confusion regarding the laws in your state. Your neighbor may say one thing and your family member may say another, but it's best to point them (and yourself!) to the correct resources and explain the most current and up-to-date information you can get your hands on. This chapter will show you how to do just that.

Why is there so much confusion in the first place? Let's go back to the 1800s…

THE HISTORY OF ILLEGAL RAINWATER HARVESTING IN THE US

In the US, the history of water rights has a complicated past. It's often said that rainwater harvesting is illegal in some areas of the US due to the complex history of various settlements, the availability of water (especially in the western regions), and how water has been distributed to major municipalities throughout the country. Water rights and laws are divided into two main categories: one set applies to the eastern states and one set applies to the western states. East of the Mississippi River, rainwater can be so abundant that residents must often deal with flooding, difficult stormwater management, and *too much* rain. For the most part, if you live in the eastern states, there are very few (if any) laws related to preventing you from harvesting and using the rain on your property. This chapter will help you double-check those local laws.

In the western states where high mountains cause rain shadow deserts behind them, bitter disputes about water rights are as old as the West, which was settled and colonized in the 1800s. (To make things even more problematic, other climate patterns also create deserts and dry valleys.) During those colonizing days, early immigrants to North America used rainwater to wash clothes. That led to people using the terms "hard water" and "soft water." When mineral-rich, "hard water" was used to wash clothes, soap reacted with the hard water and caused unwanted buildup, but when "soft water" such as rainwater was used, soap and dirt washed off easily.

Today, many states are home to large ranching and agricultural operations. These businesses have high water needs, and those requirements coupled with dry conditions and climate change have made it difficult to properly allocate water to growing populations, especially in the West and especially in the 20th century.

Lake Mead—which supplies water for 20 million people across the United States and Mexico —showing alarmingly low levels of water storage in 2021.

Western states have generally tighter restrictions on collecting rain than eastern states do. However, that's not always the case—in the state of Georgia, it's legal to collect water for non-drinking, outdoor-use only, while in Wyoming, there are no statewide restrictions on rainwater harvesting at all. Colorado and Utah have among the strictest statewide rainwater harvesting regulations in the US due to first-come, first-served downstream water rights laws passed when these drier states were first settled and farmed. This is rapidly changing; starting in 2016, Colorado has allowed residents living on a well to collect as much rainwater as they need, as long as it's off the roof of their primary residence.

It's important to learn what your local state and county (or city) laws are concerning rainwater harvesting so that you know your potential constraints as well as where to look up any changes to the laws if changes do occur.

IS RAINWATER SAFE TO DRINK?

Some states don't allow rainwater for potable use. ("Potable" means water that's safe to drink, a.k.a. drinkable.) Whether or not that is medically justified generally has to do mainly with paranoia and statutes that attempt to protect the people from themselves. Rain collected in industrial areas, areas near heavy traffic, or off of certain roof materials like asphalt, are unlikely to be safe to drink. However, in most areas—suburban yards, rural homes, homesteads, and farms—harvested rainwater that's clear and has little taste or smell is probably safe to drink and unlikely to cause any illness for most users.[1] Concerns about "acid rain" may arise. Acid rain is rain considered to be more acidic than usual due to industrial activity. If you live near an industrial site, it's not a good idea to drink your rain, though harvested rainwater near industrial sites can still be used for outdoor uses like gardening or washing a car. The acidity or alkalinity of water is measured on a scale from 0 to 14. This scale is known as the pH scale. (0 is the most acidic; 14 is the most alkaline.) Industrial-site rain that is acid rain has a pH of around 4.0,[2] whereas most rain has a slight acidity of 5.4 due to the presence of dissolved natural carbon dioxide. Acidity should not be considered toxic or "bad." Orange juice and lemonade are acidic drinks we drink happily and readily. That said, be aware that acidic water can corrode copper pipes. Fortunately, these effects can be easily mitigated using modern-day filtration systems.

Bottom line: while some rainwater can be harmful if ingested (i.e., rainwater collected near industrial sites or heavy traffic areas, or off certain roof materials like asphalt), most rainwater is safe. If you *do* live near an industrial site or heavy traffic area, the rainwater you collect may be considered "acid rain" due to the presence of pollutants. As such, restrict its use to general outdoor uses or for watering nonedible plants.

SUMMARY OF STATEWIDE RAINWATER HARVESTING REGULATIONS IN THE US

Before you begin harvesting rainwater, it's best to check with your state departments regarding the laws governing rainwater collection in your area. In the US, the relevant departments include your state's Department of Agriculture, Department of Health, and local water boards. Go to their websites and look up materials related to the keywords "rainwater harvesting" or "rainwater collection."

The US federal government does not have any laws or regulations regarding rainwater harvesting, but most states allow citizens to collect rainwater and even encourage them to do so—states like Rhode Island, Texas, and Virginia encourage residents to collect rainwater by offering a tax credit or a tax exemption for equipment purchased for rainwater harvesting. In contrast, states like Colorado and Utah have rainwater harvesting regulations in place to limit how homeowners harvest and use rainwater on their property. Each county and municipality may have additional incentives, rebates, or codes attached to rainwater harvesting on a more local level. For example, to protect their storm drains, the City of Raleigh, North Carolina offers reimbursements of up to 90% of the cost of rainwater harvesting projects. Check your city or county website for more information. In general, the laws in North America are more stringent when it comes to potable water than they are for non-potable water.

Check the table below to find out the status of rainwater harvesting in your US state. Note that while these are up-to-date regulations as of February 2022, because these laws change, you'll need to check the Rainwater Harvesting Regulations Map on the Energy.gov website to get the most up-to-date codes for your state. Rainwater harvesting regulations can be defined as granularly as at the county, city, and township level.

If you don't live in the US, research your country's regulations regarding rainwater harvesting. These tables are intended to illustrate and guide, not to provide legal advice. Always double-check your local rules and regulations!

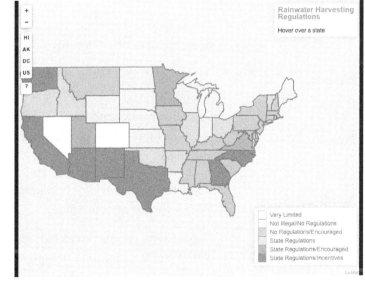

Rainwater Harvesting Regulations Map on Energy.gov

TABLE 1. REGULATED STATES IN THE US FOR RAINWATER HARVESTING

(with a Summary of Regulations)

REGULATED STATES	REGULATIONS
Arizona	Legal to collect rain on your property only, but local cities/counties provide incentives.
Arkansas	Storage is only permitted for non-potable water and if the system has been designed by professional engineer.
California	Specific system requirements and licensing criteria for system installation. Local cities/counties offer incentives.
Colorado	Allows residents to collect up to 110 gallons per household for outdoor use only. In addition, residents supplied by a well can collect unlimited rain for indoor and outdoor use.
Georgia	Outdoor non-potable use permitted only. Local cities/counties offer incentives.
Idaho	Legal to collect as long as rain has not yet entered a natural waterway.
Illinois	Non-potable use (indoor and outdoor) permitted only, and the installed system must comply with the Illinois Plumbing Code.
Minnesota	The state requires permitting and code compliance, but local cities/counties offer incentives.
Nevada	Non-potable use (indoor and outdoor) permitted only for domestic use.
New Mexico	Regulated by the state, with no requirements for outdoor use. Indoor use must meet certain quality standards. Local cities/counties offer incentives.
North Carolina	Regulated by the state for non-potable use (indoor and outdoor). Local cities/counties offer incentives.
Ohio	Regulated by the state to include specifications for system sizing and materials; allows for both non-potable and potable use.

Oklahoma	Regulated but encouraged by the state.
Oregon	Regulated by the state. For the general public, it's legal to capture rain off a roof. Local cities/counties offer incentives.
Texas	While regulated by the state, Texas offers exemption on sales taxes for rainwater harvesting equipment and has made it illegal for homeowners' associations to ban rainwater harvesting installations. Also, new state buildings in Texas now require rainwater harvesting system technology. Local cities/counties offer incentives.
Utah	Allows rainwater storage up to 2,500 gallons, with registration requirements depending on the number of containers and their storage capacity.
Virginia	Regulated by state plumbing code, but local cities/counties offer incentives.
Vermont	Regulated by the state, but no statutes regarding system specifications.
Washington	Regulated by the state, but it's completely legal to collect rain off the roof. Local cities/counties offer incentives.

TABLE 2. UNREGULATED STATES FOR RAINWATER HARVESTING IN THE US

UNREGULATED STATES			
Alabama	Kansas	Missouri	South Carolina
Alaska	Kentucky	Montana	South Dakota
Connecticut	Louisiana	Nebraska	Tennessee
Delaware	Maine	New Hampshire	Rhode Island (10% tax credit attached to rainwater storage)

Florida (local cities/counties offer incentives)	Maryland (local cities/counties offer incentives)	New Jersey	West Virginia
Hawaii	Massachusetts (offers income tax credit)	New York	Wisconsin
Indiana (local cities/counties offer incentives)	Michigan	North Dakota	Wyoming
Iowa	Mississippi	Pennsylvania	

Those living in unregulated states have the freedom to choose how they wish to use their rainwater, whether that's for gardening or laundering clothes or cooking and drinking. The only limits are a family's budget and time. That said, rainwater harvesting systems can grow and improve over time even if a household has a small budget.

For those living in regulated states, your state's regulations may not be that much of an impediment. (Texas, technically a "regulated" state, has one of the highest observable concentrations of rainwater harvesters in the US.[3]) Just do a quick Internet search on your state's website (search "[your state] laws rainwater harvesting") and your city and/or county's website to see how your state and local laws affect you. Because this step will drive any further decisions you and your family can make with respect to rainwater use in and around your home, it's important to do this research first.

ACTION GUIDE FOR STEP 1: LEARN THE LEGALITIES OF YOUR LOCALITY

Step 1.

On your Internet search engine, search "[your state] laws rainwater harvesting." Look for any and all rules and regulations that may be valid in your state and double-check that your sources are current and accurate. The Rainwater Harvesting Regulations Map on the Energy.gov website has the most recent and reputable information regarding rainwater harvesting for US residents.

Step 2.

In addition to the laws in your state, also look up the rainwater harvesting laws in your county (and if you live within city limits, your city) by looking at your county and/or city's website. If you don't find anything, there may not be any written rules or laws regarding your local area. In that case, simply follow your state's guidelines.

LEARN THE 10 BASIC COMPONENTS OF A RAINWATER HARVESTING SYSTEM

CHAPTER 2

This chapter will introduce the beginner to the essential components of every rainwater harvesting system that uses a tank. As you progress through the book, I'll explain how to choose, source, install, and maintain these components in greater detail.

10 BASIC COMPONENTS OF A TANK RAINWATER HARVESTING SYSTEM

The essential components of any rainwater harvesting system are the same—they only vary according to the size of the project and the personal preferences of the owner. Every rainwater harvesting system consists of the following:

1 **Catchment surfaces (typically a roof)**

These surfaces directly receive rainfall and supply it to the gutters and then to the storage tank. Most rainwater harvesters capture rain by collecting rain that falls on the roof on their house, but roofs over sheds, barns, or chicken coops are also fair game. Elevated, wide surfaces are key—with them, you can use large surface areas to allow gravity to bring rain down to wherever you need it to be. Rainwater harvesters in the Arizona desert even use "rain roofs." These are sheets of roofing with gutters but without a building; they're spread over the ground to collect and divert rain into underground cisterns. With the QuickRain Blueprint, you'll be harvesting rain from the roof or a catchment surface you already have. If you have an asphalt roof, don't use the rainwater for drinking! Other end-use goals are fine.

2 **Gutters and downspouts (also known as conduits or conveyance systems)**

These channels run all around a sloping roof or catchment surface and transport rainwater to a storage tank. Gutters are usually considered to be horizontal conveyance pipes and go around the roof, whereas downspouts are considered vertical conveyances and go down the length of a building. Gutters and downspouts can be made of many different weather-resistant materials, although aluminum, zinc, and galvanized steel are the most commonly available. With the QuickRain Blueprint, you will harvest rain using the gutters you already have.

3 **Screens and filters for gutters, downspouts, and tanks**

Filters remove larger suspended debris and pollutants (such as leaves and sticks) from rainwater collected from the roof. A rainwater harvesting system can have multiple filters, from screens on the gutters to filters that fit in line with downspouts to a filter on the opening of the tank itself. You can also have floating filters inside the tank to provide a final filtration step before water enters a pipe or pump. If you don't currently have gutter screens, you'll find out how to source them (and other downspout and tank screens and filters) in Chapter 5: Scale and Source Your System.

Storage tank   4

Storage tanks can be barrels, buckets, tanks, or cisterns. They hold the water once it's been diverted off your roof and gutters. Tanks are typically the first big purchase a rainwater harvester makes. They come in all shapes and sizes; the most cost-effective and accessible types are plastic and steel tanks. Where the tanks are placed depends on space availability—they can be placed aboveground next to a home, inside of a home, or even entirely underground. Some regular maintenance measures like cleaning and disinfecting are required to ensure the quality of water that's stored in the tank or cistern. With the QuickRain Blueprint, you'll place your tanks aboveground and route rainwater from your roof to your current gutter system and then straight to the storage tank.

Storage tank foundation   5

Water is heavy, much heavier than you think! A gallon of water weighs 8.3 pounds, so a full 55-gallon rain barrel will exert over 450 pounds of pressure on the ground beneath it. If the barrel is sitting on a soft patch of mud, it could sink several inches down. Magnify this problem with cisterns, and this could compromise your storage tank. Avoid unsteady foundations such as a pile of pallets! When that collapses, the person (or animal) underneath it will have a very bad day. Foundations should be as sturdy and immovable as possible. For a 55-gallon rain barrel or IBC tote, cinder blocks beneath the barrel should be fine, but for cisterns that hold over 500 gallons, consider installing a compacted gravel foundation or a concrete pad so that the cistern's weight can be securely and well distributed. More information on types of foundations, how to pick the kind that's best for you, and how to install one will be covered in subsequent chapters.

Storage tank vent   6

Your tank or cistern is always filled, either with air or water or both. Your barrel/tank should come with a lid to keep unwanted debris and animals out, but it should also allow air to pass in and out of the tank as water empties or fills. The storage tank vent may be a separate screened opening on the tank that can double as an overflow outlet. If the lid is on too tightly and the tank's water empties without enough air entering and neutralizing the pressure in the tank, the tank can implode. (Or, if the opposite happens, the tank can *explode*!)

7

Storage tank overflow

It's not a matter of *if* your storage tank overflows—it's *when*. Plan for it! Your storage tank will need a pipe that routes any overflowing water away from any building foundations and hopefully to somewhere useful, like a garden bed or a tree that needs to be watered. Wherever the overflow outlet level on the tank is, that's the maximum level the tank will hold.

8

Storage tank outlet

This is where you'll access the water that's at the bottom of the tank. The outlet can have a spigot installed at the outlet, or it can have a pipe that's connected to a pump or a finer filtration system. You can attach a hose to the faucet to bring the water from the point of collection to where you need it.

9

First-flush system (optional)

This system ensures that the first batch of rainwater—which is usually the most contaminated—is flushed out and not harvested into the system. For most outdoor uses, this diverter is not necessary. But for those who want to water edible plants, water livestock, or bathe with rainwater, first-flush diverters can be a great option to use instead of using potable water, as doing so can save you the energy of treating raw rainwater until it's potable. However, first-flush diverters also have downsides, chiefly that they need to be cleaned and inspected regularly to prevent bacteria from colonizing the water in the pipe. But again, depending on your location and your rainwater's end-use goals, you may not need first-flush diverters. (In Chapter 5: Scale and Source Your System, you can decide if first-flush diverters are worth using in your particular situation.) You can use the water emptying out of the first-flush diverter as a resource by planting a tree or putting a trellis near the first-flush diverter. Then whatever grows there will be watered with the first-flush water and can climb up your tank to beautify it.

10

Pumps and indoor use filtration and treatment (optional)

Pumps are not necessary, especially for many outdoor systems that just need to water a garden. Pumps *are* necessary when water pressure is critical and that pressure can't be created by height. If you're watering a garden, you can keep complexity and costs low (especially at first) by filling a watering can from the rain tank rather than using a pump. To increase water pressure at the point of use without a pump, use a taller tank or elevate the tank off the ground. If you're still dissatisfied with your water pressure, then consider purchasing a pump. You'll also need a pump if you wish to use a hose or a pipe to bring rainwater from the tank to anywhere else that's above the tank's grade. You will very likely need a pump if you plan to use rainwater indoors for faucets and appliances. Pumps run the gamut in complexity (from simple to very complicated) and affordability (from cost-effective to very expensive). If a pump makes it easier to use your collected rainwater, then it can encourage you to use the water in your tank. If you plan to use the water indoors, you will also need finer filtration (that is, between 20 and 5 microns), and if you plan to drink the water, you will also need to remove any potential disease-causing microbes by using one of various methods of treating water for potability (some of those methods include a countertop filtration system like Berkey, UV disinfection, or chlorine disinfection). More information on deciding which pump, filtration, and type of treatment could be most useful to you will be covered in subsequent chapters.

While the focus of this book is on roof catchments and tank storage systems, earthworks deserve a brief mention for their ability to utilize rain in passive and efficient ways. They're especially useful for gardeners and farmers who plan to use rain to water ground-level plant beds as well as for those looking for flood and erosion control around their buildings and yard.

INTRODUCTION TO EARTHWORKS

Earthworks are features of earth that have been shaped in ways that enable them to move rainwater when it reaches the ground. Earthworks deserve a mention because of how effective they are at watering plants and irrigating the land with little to no maintenance. Here are examples of three simple earthwork technologies for beginners: French drains, berms, and swales.

1. **French drains:** A French drain is a long, narrow, gravel-lined trench that has a pipe underneath it for water redirection. It's analogous to a gutter, but it's used for groundwater rather than rainfall. People use French drains to address pooling groundwater near foundations that can cause cracked foundations and other problems. If you have water pooling near your foundation, these could be a great option to look into! Also, if you're considering hiring a foundation professional, ask about French drains as a water diversion solution.

2. **Berms:** A berm is a raised mound that retains water, allowing the water to soak into the soil. They are useful for areas that don't get enough water or for garden beds that could benefit from long watering times.

3. **Swales:** A swale is a shallow ditch with sloping sides that intercepts water runoff, serving as a channel for water. Swales control stormwater running off hillsides and can direct it to places where you need it, like a tree or garden bed. Swales are generally simple to install—you can use a simple shovel or a mini-trencher to carve channels into a hillside or along your yard that will then allow water to be collected and redirected. Swales are usually more suitable for irrigation and water diversion in yards, whereas French drains are used more often to address foundation problems and basement flooding and to control water in and around buildings and homes.

Other kinds of earthworks include rain gardens and terraces. If you believe earthworks are your solution, check out *Rainwater Harvesting for Drylands and Beyond: Volume 1* and *Rainwater Harvesting for Drylands and Beyond: Volume 2* by Brad Lancaster or *The Permaculture Earthworks Handbook* by Douglas Barnes.

ACTION GUIDE FOR STEP 2:
LEARN THE 10 BASIC COMPONENTS OF A RAINWATER HARVESTING SYSTEM

Step 1:

Become familiar with the 10 basic components of a tank-based rainwater harvesting system. By doing this, you will be aware of any components you may be missing as you source the materials for your system and install your system.

Step 2:

Decide if a tank-based rainwater harvesting system will fully meet your needs or if your property could benefit from earthworks. If your property *could* benefit from earthworks, learn more about them from the books recommended in this chapter as well as online. Earthworks can be simple and fun to experiment with—just get a shovel and start digging!

CHAPTER **3**

DECIDE ON THE END-USE GOALS FOR YOUR RAINWATER |

Your family's needs, your budget, and your local laws governing rainwater harvesting will determine what you can do with rainwater now and what you can potentially do with it in the near future. Whatever you decide to do with it (e.g., if you just need it to water your garden vs. you need it as a primary or secondary water source) will drastically affect the size, complexity, and cost of your system. In this chapter, you'll determine which end-use goals could be right for you right now. You'll also learn how rainwater can be used for other end-use goals.

The uses of harvested rainwater fit into three categories that become more complex and costly with each use. These categories are outdoor non-potable use, indoor non-potable use, and indoor potable use.

OUTDOOR NON-POTABLE USE

Outdoor non-potable water typically consists of raw rainwater harvested straight from a roof into a barrel or storage tank. It passes through outdoor gutters and straight to applications that don't require fine filtration or treatment. That said, some filtration may occur to keep larger debris such as leaves and sticks out of the water. Because outdoor non-potable rainwater uses the least amount of filtration, it's considered the most accessible and inexpensive way to use rainwater.

Watering your plants with rainwater is one great way to use outdoor non-potable rainwater.

Outdoor non-potable systems can range quite a bit in price depending on how big the tanks are. Most backyard and homestead systems for outdoor non-potable use cost between $100 (for a glorified rain barrel) and $4,500 (for a 3,500-gallon tank with a pump). Some uses for outdoor non-potable rainwater include:

Gardening and irrigation: You can use rainwater in watering cans to water plants by hand, or you can attach a rainwater storage tank directly to an automatic irrigation system. When it comes to watering edible plants with rainwater, there are two schools of thought. One is to use non-potable water and water just the soil instead of the plant so that the non-potable water isn't directly ingested. The other is to treat rainwater to make it potable (which is not always the easiest or cheapest thing to do) and water the plant itself. I recommend collecting as much rainwater as you can—whether it's potable or not—and then watering the soil with this free resource.

Watering animals: Many farmers, ranchers, and homesteaders water outdoor animals with filtered, nontreated rainwater. As long as the rainwater is harvested using nontoxic materials (more on materials in Chapter 4, Section 2: Not All Materials Are Safe), large debris has been filtered from the rainwater, and the system is properly inspected and maintained, rainwater is a pure, soft, natural, and sustainable way to water backyard pets and livestock. Another school of thought maintains that watering pets and livestock should only be done using potable water. This is a personal choice, but in all cases, be sure to properly maintain your rainwater harvesting system anyway, as this keeps the quality of the rainwater high without vigorous water treatment. (Maintenance is covered in Chapter 7: Maintain Your System.)

Flood control and erosion prevention: Rainwater that enters your garage, basement, or home does so due to several factors, including the improper diversion of rainwater away from your property. Properly seal your buildings from water, of course, but also inspect your property's slope and terrain. Management of flooding and erosion is typically outside the realm of storage tanks and more in the realm of earthworks and foundation waterproofing, but it still deserves a mention here. Earthworks such as berms, swales, and French drains divert water passively, and building them will prevent stormwater damage to your home via that diversion. You can find more information on flood control and erosion prevention in *Roofing, Flashing, and Waterproofing (For Pros By Pros)* by the editors of *Fine Homebuilding*. This book offers an affordable way to inform yourself about your situation before hiring a waterproofing professional.

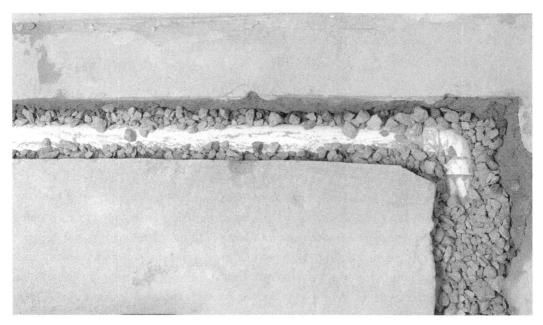

A French drain in action, directing water away from a building's foundations.

Composting: Water is essential for the proper decomposition of your compost pile! Use rainwater to hydrate the microorganisms and worms that are decomposing your compost. This will also prevent your compost from overheating.

Car/house washing: Use free rainwater to wash your car and home! Not only will you be using a free resource for a resource-intensive operation, but rainwater is soft and will leave a streak-free finish. With no calcium or chlorine content, it's ideal for washing floors and windows because it won't leave spots or streaks, plus rainwater suds more easily than hard water does. If you currently have issues with hard water, rainwater may be a solution.

Car and house washing can use a large volume of water. Save money by using free rainwater that will leave a streak-free finish!

Fire protection: A rainwater catchment system with a large storage tank could give you some extra protection from fires, especially if you live in a region that's prone to wildfires. It's also essential to install a good pump so that you can access the water quickly if needed. If fire protection is suitable for your home, take a good look into Aquamate (aquamatetanks.com) or Pioneer Water Tanks (pioneerwatertanksamerica.com). They service Australian homes prone to bush fires and Californian homes prone to wildfires by installing fire-resistant tanks.

INDOOR NON-POTABLE USE

Indoor non-potable rainwater typically consists of outdoor non-potable rainwater that's been more finely filtered from sediment that can clog pipes and machines. Filters for indoor non-potable use are typically 100 microns or less[1] to keep pumps, indoor pipes, and appliances free of particulates. Indoor non-potable systems require as much filtration as an outdoor non-potable system does, plus you'll need extra finer filtration as the water exits the rainwater tank into a pump and then into your home's pipes and appliances.

Most residential indoor non-potable rainwater harvesting systems cost between $300 (for a 55-gallon rain barrel with fine filtration and a tiny pump) to $8,000 (for a 5,000-gallon tank with fine filtration and a very nice pump). Examples of what these systems can look like will be illustrated in subsequent chapters. Some uses for indoor non-potable rainwater include:

Toilet flushing: Toilets account for nearly 30% of a home's water consumption. To flush a toilet with rainwater, filter rainwater with a fine sediment filter (around 100 – 20 microns) to remove debris and particles that might otherwise block pipes and valves. More on where these filters can be sourced is covered in Chapter 5: Scale and Source Your System.

Washing clothes: Harvested rainwater is great for washing clothes, particularly in hard-water areas. Because rainwater is naturally soft, it will use less detergent and will prevent the buildup of limescale, thus prolonging the life of your washing machine.

Bathing: Showering is a fine way to use rainwater, as it typically accounts for 20% of indoor water usage. Because water can potentially enter your nose and mouth as you shower, some people argue that it's best to only shower with potable rainwater. However, plenty of people argue otherwise, and millions of people around the world shower with non-potable water. If you don't plan to use water for potable reasons at all, then it's probably not worth treating rainwater until it's potable just to bathe with it. If you live in a rural or suburban area, it's likely your rainwater quality will be high, and employing a first-flush diverter can help you use only the highest-quality rain runoff from your roof.

Take a bath with rainwater without having to treat it for potability.

INDOOR POTABLE USE

If you have an asphalt roof or live near a high-traffic or industrial area, don't use your rainwater for potable uses, even if it's treated for bacteria! Toxins in the water can be difficult to remove, even with filtration. Also, **an asphalt roof is considered a toxic material**. While water collected off an asphalt roof could be suitable for outdoor use, it's not suitable for ingesting.

Indoor potable water needs to be filtered to remove even extremely fine sediments (5 – 20 microns). It also needs to be disinfected to prevent contamination from microbes and pathogens such as *E. coli* and *Giardia intestinalis*. Disinfection can occur via several different methods, including but not limited to adding chloramine (chlorine), radiating with ultraviolet (UV) light, boiling, or pouring through a countertop potable water filter such as a Berkey unit. If you live near an industrial area or an urban area where pollutants or toxins in the air could easily enter your rainwater, don't drink the rainwater, not even if it's been disinfected. That said, most people who live in rural, suburban, and even most urban areas should feel confident about drinking their rainwater once it's been treated properly. More information on potable water treatment can be found in Chapter 5: Scale and Source Your System.

Treatment for potable water use can range from inexpensive (some use only chlorine bleach) to very expensive (some use UV light or microfiltration systems that can cost up to roughly $2,500). Potable water treatments with a low barrier to entry (e.g., countertop filters) are great starting points, but they may end up costing more per gallon of water purified over the long run. Most residential indoor potable rainwater harvesting systems cost between $800 (for a 500-gallon rain tank with fine filtration, a pump, and potable water treatment) to $8,000+ (for a 5,000-gallon tank with fine filtration, a pump, and high-end potable water treatment). Some uses for indoor potable rainwater include:

Watering edible plants: Some people recommend watering edible plants (e.g., vegetables, fruits, herbs, etc.) only with potable water in an attempt to minimize the body's exposure to harmful bacteria. This is a personal choice. Note that edible plants can be watered with non-potable water via watering the soil only. Also note that soil already contains *E.coli* and other bacteria anyway.

Watering animals: If you're used to watering pets with municipal, chlorinated water only, you may decide to water backyard pets and livestock with potable rainwater only. And if you decide to treat rainwater for indoor potable use for you and your family, using the same water for your pets is a great option. Do note that animals are often watered with water that hasn't been disinfected as rigorously as it typically is for humans. Rainwater that's harvested using non-asphalt roofs, properly filtered for debris and sediment, *and* periodically inspected can be considered pure and of high quality, even for pets and livestock.

Bathing: If you and your family don't want to risk bathing with non-potable water, treat your water for potability and find peace of mind every time you bathe with rainwater. People with sensitive skin can benefit from bathing from potable water.

Washing dishes: The water you use for washing dishes is just as important as the water you use for cooking and drinking. And remember, rainwater is soft, making it easy to lather. It will give your dishware a squeaky-clean finish!

Cooking and drinking: Rainwater is safe for both cooking and drinking. All you need to do is take the preventive measures necessary to filter and purify the water to make it potable. More information on treating water for potability can be found in Chapter 5: Scale and Source Your System.

Best practices encourage washing dishes with potable water to avoid the risk of ingesting harmful bacteria.

TABLE 3. SUMMARY OF RAINWATER USES

Increasing filtration/treatment required

Outdoor Non-Potable Use	Indoor Non-Potable Use	Indoor Potable Use
Gardening and irrigation, watering edible plants (option to be potable)	Toilet flushing	Watering edible plants
Watering animals (option to be potable)	Washing clothes	Watering animals
Flood control and erosion prevention	Bathing (option to be potable)	Bathing
Composting		Washing dishes
Car/house washing		Cooking and drinking
Fire protection		

FACTORS THAT AFFECT THE COST OF RAINWATER HARVESTING

- Startup Costs of Rainwater Tank Systems

The cost of a given storage tank depends on its size and the materials it's made of—the bigger the tank, the more costly the tank is. Cheaper tanks are made of plastic (like a high-density polyethylene), while more expensive ones are made of galvanized steel and stainless steel. Other tanks are made of fiberglass and concrete. Fiberglass and concrete tanks are more difficult to source for residential systems due to their higher prices and lower demand. You can read more about tank types in Chapter 5: Scale and Source Your System.

More sophisticated designs of tanks also carry different considerations in terms of cost and time:

- You spend nothing when you use free/recycled barrels and the roof and gutter system you already have. Installing a rain barrel need not take longer than a Saturday afternoon. For a mid-tier outdoor system, expect to spend around $1,000 – $4,500 for a 500-gallon system (which includes the tank, filters, pump, and pipes) and about 10 – 12 hours installing the system. For more information on installation, see Chapter 6: Install Your System.

- The highest tier of rainwater harvesting systems can cost $12,000 – $20,000+ and take several days to install. The largest and most complex systems use professionals to install large (100,000+ gallons!) tanks with complex piping systems over a large acreage. For most backyard and homestead rainwater harvesting systems, these systems are not necessary, but describing them provides context to what's available.

Rainwater harvesting systems can also be upgraded over time. You may start using the roof and gutters you have now, but then you may want to upgrade your roof and gutters later. As an example, you can decide to change your roof to a metal roof to create a higher catchment efficiency, or you might want to purchase seamless gutters to maximize the amount of rain you can catch. You may choose to start with a 100-gallon tank now, but you may want to increase the size of your tank as you find more uses for rainwater over time. Depending on their quality, replacing these components can cost just a few bucks to several hundred dollars to several thousand dollars. You can use what you already have or upgrade it or change it! I recommend starting to harvest rain with what you have now and the components you can find at a local hardware store. Learn how your current components affect your rain catchment and rainwater quality. Only then will you have the necessary information to decide if you need to stick with what you have (and save money) or if it's necessary to change things up and spend money on upgrades.

- Costs Over Time for Rainwater Tank Systems

Your system's costs over time will factor in maintenance and upgrading expenses. Maintenance includes cleaning out gutters, roofs, filters, and tanks, as well as fixing leaks or breaks. Upgrading includes adding new features to existing devices or upgrading the materials or components of your system to higher-quality parts.

If you choose to do all maintenance and upgrades yourself, expect the costs of maintenance to include cleaning and securing gutters, cleaning out filters, fixing leaks, and possibly replacing parts. While the costs vary significantly depending on the maintenance required, expect to pay around $100 – $500 per/year post-installation for maintenance (not accounting for catastrophic failures or replacements). Budget in 5 – 10 hours/year for basic maintenance. If you find yourself with little time or you encounter catastrophic failures in your system, consider hiring a landscaper who specializes in installing and maintaining rainwater harvesting systems or a local professional rainwater harvesting company (search online for "rainwater harvesting professionals near me"). These professionals can help you fix large-scale failures such as tanks settling or imploding, flooding, etc. But don't worry— catastrophic failure is rare! If you properly maintain your system, you can avoid large-scale failures. More about this in Chapter 7: Maintain Your System.

Rainwater harvesting or landscaping professionals cost about $45 – $75/hour. Some offer year-round maintenance packages of around $1,000/year. These packages include full inspections, top-to-bottom maintenance and replacement of parts, water quality testing, and even winterization and can involve up 50+ hours/year. You likely won't reach this level of maintenance unless your system is unusually large or complex or if catastrophic failures do occur and the cleanup is too much to handle yourself. For the systems described in this book, you can do all or almost all of the maintenance yourself.

Use Table 4 to estimate the startup and maintenance costs (in terms of both money and time) of a rainwater harvesting system, whether it's earthworks, a rain storage tank, or both.

TABLE 4. STARTUP AND MAINTENANCE COSTS FOR RAINWATER TANK SYSTEMS

	Low-cost system in dollars (USD)	Average cost for a residential-scale system (USD)	High-cost system in dollars (USD)	Low-cost system in time	Average cost in time for a residential-scale system	High-cost system in time
Installation of water tanks	$0 – $100	$1,000 – $4,500	$12,000 – 20,000+	A few minutes to 1 hour	5 – 12 hours	50+ hours
Maintenance and upgrades of water tanks over time	$0 – $50/year	$100 – $500/year	$1,000 – $5,000/year	1 – 5 hours/year	5 – 10 hours/year	50+ hours/year

ACTION GUIDE FOR STEP 3: DECIDE YOUR RAINWATER'S END-USE GOALS

Step 1:
Determine what your water needs are for your family and your home. Using the information from your state and local laws as well as what you've determined to be your family's water needs, decide on the end-use goals for your rainwater: outdoor non-potable, indoor non-potable, indoor potable, or a mix of the three.

Step 2:
Determine a ballpark budget for your rainwater needs based on your end-use goals. Remember that rainwater harvesting systems can be small and inexpensive for now but can grow in size, complexity, storage capacity, and cost in accordance with what your future needs and curiosity dictate.

LEARN THE BASICS OF RAINWATER HARVESTING SAFETY AND USE

N ow that you've learned about your local laws, the basic components of rainwater harvesting systems, and how you intend to use your rainwater, let's learn the basics of rainwater harvesting safety. Safety guidelines with respect to rainwater are similar to those of water in and around the home in general: keep water away from your foundations, make sure your storage tank doesn't become a breeding ground for unwanted pests, and be prepared for your tank to overflow, because it will. Keep these safety basics in mind, and you will protect yourself, your family, and your home with ease.

Basic rainwater harvesting safety measures are sufficient for most people who are collecting rain for outdoor non-potable use or those who are just collecting water for their garden. However, if you intend to use your rain for indoor use, whether potable or not, you'll want to have an idea of how your climate, your roof, and your household needs will affect your rainwater supply. That's where the basics of rainwater harvesting *use* come in. This chapter will walk you through learning how your location's rainfall patterns can prepare you to capture as much rain as possible, how much rain your roof can bring into your system, and how to know if that will be enough so that you can prepare a backup source of water if need be. Knowing all of this is the key to harvesting rainwater with confidence. Let's go!

Rainwater Harvesting Safety Rules

1. Don't mix water and your foundation
2. Not all materials are safe
3. Protect your tank from sun, insects, and animals
4. Plan for overflow
5. Water is heavy!
6. Ensure end-use water quality
7. "Low maintenance" does not mean "no maintenance"

Rainwater Harvesting Uses Rules

8. Understand your climate
9. Understand your catchment area
10. Understand your water needs

1. DON'T MIX WATER AND YOUR FOUNDATION

Nothing will stop your rainwater harvesting dead in its tracks faster than a compromised, leaky foundation. Have you ever inspected your gutters and watched the water flow off of your property when it rains? Does it properly flow away from your foundations? Ultimately, water pooling near your foundation can build up next to the foundation's walls, which can lead to water seeping into your home and impairing the integrity of your home's foundation. Any cracks in the foundation can be made worse if there's high water pressure next to the foundation's walls. Along with preventing water from pooling near your foundations, efficiently diverting water away from your foundation with your gutters and downspouts is your best defense against water causing damage to your foundation.

Signs of a leaky foundation, which should be addressed as soon as possible. Do not make this situation worse with overflowing rain tanks positioned near your home!

Clean and inspect your gutters up to twice a year (and more often during the rainy season). Don't just do this to keep your rainwater quality high—also do it to prevent water from damaging your home. Overflow pipes from a rain storage tank are extremely important for this reason because they divert the overflow away from your home, preferably to somewhere useful that can easily absorb the water (like a sunken basin). Overflow pipes also keep overflows from flowing to neighbors' homes.

If you have any concerns at all about the ability of your foundation or basement to keep water out, solve your foundation problems *before* playing with gutters and water diversion on your property. One DIY way to address this issue is to add soil (preferably a high-density soil like clay) near and around your foundation to create a sloping-away effect. This helps channel water away from your foundation/basement. You can also purchase waterproofing membrane or paint at your hardware store to insert into foundation or basement cracks to seal them. If you still have foundation issues, postpone rainwater harvesting until you feel confident that potential overflows won't further damage your home. You need to take the integrity of your foundation seriously! If you doubt your own expertise in addressing this, it may be time to call a foundation or waterproofing professional.[1]

If your foundation is good to go, then you're good to start harvesting rainwater. Let's continue!

2. NOT ALL MATERIALS ARE SAFE

Many studies have been done on the effects that roofs, gutters, pipes, and storage tanks have on water quality in terms of their materials. In short, different materials affect water quality in different ways. Ultimately, you must decide whether the materials you currently have or will acquire for rainwater collection and storage are suitable for you and your family. While many materials are safe, you'll need to be careful with a small number of them.

Roof, Gutter, and Downspout Materials

Asphalt roofs are suitable for collecting rainwater for outdoor use, but there are concerns about using asphalt surfaces to collect rainwater for indoor use and potable use. As asphalt is considered a toxic material, it is not advisable to drink or use water in the home that comes from an asphalt roof. The only acceptable use for this water in the home is for flushing toilets.

An asphalt roof, one of the most common roofing materials. Rainwater collected off asphalt can be suitable for non-potable uses.

As for gutters, almost all gutter materials are safe to use, but be wary of vinyl gutters. They can degrade in sunlight and leach carcinogens into the water. The installation guides in this book use PVC pipes because of how accessible and inexpensive they are; however, know that they will degrade and become brittle in sunlight as well. Also, they too can leach carcinogens into the water as they degrade. Replace PVC pipes about yearly or on an ad-hoc basis if you notice cracks in the PVC pipes.

An example of vinyl gutters, which are inexpensive but can degrade more quickly than other gutter materials.

If you have the option of choosing your gutter and downspout materials and you have a suitable budget, choose gutters and downspouts made of metal such as aluminum or copper. While these may be more expensive, you won't need to replace them nearly as often, plus these materials don't carry the same risks as PVC materials do.

The most common and easily found pipes that offer the greatest range of sizes are PVC Schedule 40 pipes (the Schedule refers to the thickness of the pipe). If you plan to use PVC Schedule 40 pipes for your rainwater conveyance, before installing the pipes, paint the pipes with a few coats of paint to reduce their exposure to sunlight and prolong their lifespan. Also, plan to replace them after a year or so. Alternatives to PVC Schedule 40 pipes that can be found at hardware stores are PVC Schedule 80, HDPE, and CPVC pipes. While more difficult to source, especially of as many sizes as PVC Schedule 40, these kinds of pipes can last a longer time and with fewer risks.

Storage Tank Materials

The storage tanks recommended in this book are made of food-grade HDPE (high-density polyethylene) and food-grade metal (galvanized or stainless steel, usually with a potable water liner). As long as they're rated for potability under the NSF 61 standard (look at the manufacturer's description of the tank to ascertain this), all of these tanks are suitable for all uses, including potable use. Most tanks available online are rated as being potable under the NSF 61 standard. Consider a nonplastic tank if you have concerns about using plastic. Different materials for tanks are covered in Chapter 5: Scale and Source Your System.

3. PROTECT YOUR TANK FROM SUN, INSECTS, AND ANIMALS

You can keep your rainwater quality as high as possible *without* spending more on back-end filtration and treatment by keeping your rain tank sufficiently sheltered from the sun (use an opaque tank), screened from insects, and tightly covered to prevent animals or children from accidentally finding their way into the tank. Choosing opaque tanks not only prevents algae growth—which can be a problem for filtration systems—it also prevents your rain tank from turning into an energy-rich pond teeming with microorganisms. Most off-the-shelf rain barrels and tanks are usually opaque, although you might come across a translucent tank like a white IBC (intermediate bulk container) tote. If you have the option, purchase black IBC totes instead. If those are unavailable (or unavailable in the quality you need), purchase white totes and paint them until they're opaque. Or you can wrap them in opaque plastic or fabric sheeting like you would a present.

Be wary of using an IBC tote without painting or covering it!
Transparent totes can let in sunlight and promote algae growth.

Keep the tank sufficiently screened from insects and mosquitos by using screens with holes that are 1/16" or smaller in diameter. Almost all off-the-shelf tanks come with screens suited for this purpose, but if you're fabricating your own screens, keep that principle in mind. Window screens and pantyhose are common DIY materials that fit this criteria, but you can search online for "rain tank screens" or "rain tank filters" to find manufacturers who specifically make filters to prevent insect infestation. Every time you clean your gutters, gutter screens, and downspout filters, double-check to see if there's any standing water in the gutters or overflow pipes. If so, consider cleaning the gutters more often and/or changing the slope of any pipes to allow them to better release standing water.

If your rain tank has a removable cover or has a filter that can be removed, consider how difficult it is to remove. Raccoons are not idiots. Those little fingers love to find a free water source if they can. Don't let them! Tightly screw on covers. Attach filters to the tank with screws. Keep all pipes screened, even overflow pipes. Animals can drown in the tank, which will compromise your water quality. If you do find an animal carcass in the water, you will likely need to dump out the entire tank and disinfect the inside. Don't worry—this is not very common, especially not if the tank is sealed properly. Instructions for tank disinfection are in Chapter 7: Maintain Your System.

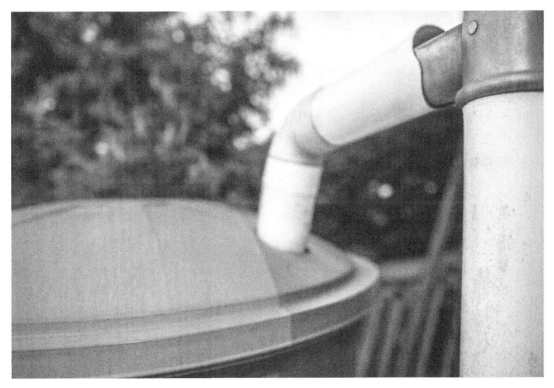

Consider how tight your tank's lid is and make sure it can't be easily removed by an animal or a child.

Mosquito Prevention

Despite your best efforts to keep your tank screened from mosquitos, it can be tricky to fully keep them out. Here are several mosquito prevention practices for keeping your yard/property as mosquito-free as possible:

1. Clean and empty any water-holding containers in your garden or yard.

2. Keep gutters clean and free of standing water.

3. Keep mosquitos out of pipes by using a screen on every inlet of every pipe (e.g., as rain travels from your gutter to your inlet pipe, use a gutter downspout screen to keep mosquitos out of the inlet pipe). Also consider installing pipe fittings called vent flaps or "Mozzie Stoppas." These are popular in Australia and available in the US on RainHarvest.com. They attach to the ends of pipes and have mosquito-proof screens. An example of vent flaps in action is covered in Chapter 6: Install Your System.

4. Use a first-flush diverter to keep mosquito eggs out of the rain storage tank. First-flush diverters divert the first flush of rainwater off the roof so that the dirtiest water doesn't end up in the tank. First-flush diverters need diligent upkeep, but they are useful in some scenarios. (See Chapter 5: Scale and Source Your System to decide if a first-flush diverter is worth it for your scenario.)

5. Keep an eye on all screens during your regular maintenance schedule (see Chapter 7: Maintain Your System) to make sure that all screens are intact.

6. Place Mosquito Dunks in the water. These are tabs of bacteria that kill mosquito larvae. They're available at hardware stores or online.

Mosquito Dunks are placed directly in the storage tank; these tabs kill mosquito larvae when they hatch.

4. PLAN FOR OVERFLOW

Installing your rain tank in a sunken basin, planting a rain garden near your rain tank, and installing swales next to your rain tank so that water can flow to nearby vegetation are all examples of using overflowing rainwater from a tank productively. An unexpectedly large rain event is never too far away, and water can be destructive if not managed properly.

As you start to consider where the rain tank(s) will be located, consider what would happen if a lot of rain pooled in that area. Would it be troublesome for your foundation if the rain tank overflowed too close to your home? Is there going to be so much standing water that you'll have created mosquito breeding grounds? Is the soil in that area so dry that the water could wash away the topsoil, causing unwanted erosion? A technique commonly used in permaculture circles to prevent these difficulties is to create a "living sponge" near a rain tank. A living sponge is a net of vegetation, swales, and berms that slows water down as it flows along the ground, allowing the water to fully infiltrate the soil to nourish the land, recharge the groundwater, and prevent erosion. You can create a living sponge near your rain tank by digging a slightly sunken basin for your rain tank's foundation, installing the tank's foundation in the sunken basin, and planting trees and other vegetation in the basin while also cutting an outlet from the sunken basin to route overflows to other parts of your yard where it might be needed. You can route the overflow pipe from the overflow outlet on the tank straight into your living sponge or to other parts of your yard.

A storage tank placed directly in a sunken basin with its overflow pipe routed straight to plants.

Also consider the danger of water pooling near your foundation if the tank overflows! If a sunken basin or other earthworks cannot be installed for any reason, route the overflow outlet far away from the building's foundation. Also point the pipe downhill and away from any neighbors' foundations, too. If you manage the overflow water as a resource, then rainwater will be seen as the gift it is rather than as a nuisance.

5. WATER IS HEAVY!

While I've already said this, here it goes again: **don't underestimate the weight of water!** Full 55-gallon rain barrels weigh over 450 pounds; full 275-gallon IBC totes weigh over 2,000 pounds; full 5,000-gallon tanks weigh over 41,500 pounds!

Before you install your rain tank, make observations about the quality of the soil where you plan to install your tank. Is the soil firm and consolidated, or is it loose and soft? Install the best, most stable, most solid foundation you can afford. (More information about the different types of foundations will be explained in Chapter 5: Scale and Source Your System.) Be wary of the danger of tanks filled with rainwater tipping over onto the wrong person at the wrong time. While that's unlikely, the more elevated a tank is, the higher the risk it carries of causing serious damage if it falls or tips over. The quality of the topsoil near your home may be loose and could cave in if too much weight is placed on top of it. Every type of soil is different! That means you need to **always inspect the compactness of the topsoil before installing the rain barrel or tank**. For tanks about the size of a 275-gallon IBC tote or bigger, if the topsoil is not firm or consolidated, dig out the first layer until you reach compacted soil. Then fill in the hole with compactable gravel and compact the gravel with a tamper tool or a vibrating plate compactor.[2] Only then install a compacted gravel base or concrete base on that spot. Tamp the base down with a tamper tool or compactor tool.

An IBC tote placed on a precarious foundation of wooden pallets can be an accident waiting to happen.

Slimline tanks are notable for fitting nicely alongside to buildings, but they're more prone to tipping over than their cylindrical counterparts are. Prevent injuries by affixing slimline tanks to the adjacent building with a tie.

6. ENSURE END-USE WATER QUALITY

You'll need to match your water filtration and treatment needs to your end-use goals. For outdoor-only uses, the filtration needs are simply gutters and downspout filters plus some filters on the tank and perhaps inside the tank. For indoor non-potable uses, you'll need as much filtration for outdoor use plus filtration for finer sediment. For indoor potable uses, you'll need finer filtration than you would for indoor non-potable water plus treatment to remove disease-causing microorganisms.

Keep your filtration and treatment as simple as your end-use goals necessitate, but also be careful *not* to skimp on filtration and treatment needs if you decide to use rainwater inside the home. Exactly which types of filtration you'll need and where to purchase them will be explained in Chapter 5: Scale and Source Your System.

7. "LOW MAINTENANCE" DOES NOT MEAN "NO MAINTENANCE"

Once rainwater harvesting systems are set up, they mostly run on their own. However, "low maintenance" does *not* mean "no maintenance"! If you take care of your system, your system will take care of you. Keep downspouts and screens free of leaves and debris and clean your gutters at least twice per year. Do visual inspections of your system on a weekly to monthly basis to ensure that there are no leaks. Before freezing weather hits, unplug and drain pumps and perform other winterization tasks. Full information about maintenance and winterization will be covered in Chapter 7: Maintain Your System.

It is *your* responsibility to maintain your system! After all, when you pay a municipal water bill, maintenance is included in that bill. When you aren't paying those bills because you're creating your own water supply, those maintenance costs and tasks fall on you. Keep yourself and your family safe by staying on top of your maintenance responsibilities.

8. UNDERSTAND YOUR CLIMATE

Understanding the general seasonality of the rain in your area will give you a full-circle understanding of when it's best to prepare for the rainy season, when it's best to conserve water during the dry season, and roughly what size tank you and your family may need to last through the drier months. In addition, if you live in a very cold climate where water in aboveground tanks may freeze and damage your tank, you'll need to time your winterization preparations in accordance with when those deep freezes will happen.

Understand Your Rainfall

Rainfall data can be captured at different levels, ranging from a very granular one to a less granular one: every day, every week, every month, and every year. For those who capture their rain data themselves, it is most helpful to know the average amount of rainfall *per month* over the course of several years. (The more years, the better! This increases the likelihood that the average will correctly predict your actual monthly rainfall.) Data more granular than monthly can vary greatly, and data less granular than monthly can be difficult to use for predicting if your rainwater supply will meet your family's water needs. Monthly rainfall data is available online, especially for residents of the US. I'll explain how to acquire this data shortly. First of all, know that the amount of rainfall is measured as a unit of length. If you leave a flower pot outside during a rain event and find it full afterwards, you probably had a torrential downpour. But if you collected all of the rain that fell on a mall parking lot and that amount filled up the same flower pot to the same level, then it hardly rained at all. This means that the amount of rain that fell is better measured as the total volume of rain that fell over the entire area it fell over. When you divide volume by area, you get a one-dimensional unit of measure— i.e., length.

Measure your rainfall with a rain gauge that displays units of length. (For the US, this is in inches.)

Rainfall is measured in inches in the US, and in millimeters elsewhere. Inexpensive rain gauges can be purchased online or from hardware stores, and it's a good idea to keep one on hand to have an idea of how much water will enter your rainwater harvesting system when it rains. (I'll provide a calculation for how to do this in the next section.) In addition to that, you'll want to acquire data on the average rainfall per month in your area for the past 20 or so years. The more years of data you can get, the better! Here's how to get your hands on that information for residents of the US:

1. Go to the National Weather Service website at weather.gov and search for your city or ZIP code. If you can't find data for your specific location, try a location nearby. Each locality collects data via a local forecast office, so variations in report content and type may occur. Still, each locality should have a report that can provide past rainfall data.

2. When you find your location, find your local climate data by looking for reports labeled "Temperature/Rainfall Records" or "Climate and Past Weather" or something similar.

3. Look for reports that provide past rainfall data, ideally monthly and ideally within the last 20 years. You'll likely receive a table of data.

4. Take a moment to observe which months of the year on average receive more rainfall and which receive less. Look at the average rainfall per month and per year. "Mean" indicates the average rainfall for that month, "max" indicates the most rainfall that month received in the past 20 years, and "min" indicates the least rainfall that month received in the past 20 years. I've included some snapshots of a sample area (Asheville, North Carolina) for you to review.

🔒 weather.gov   ⬆ ☆

NATIONAL WEATHER SERVICE

HOME   FORECAST ▾   PAST WEATHER ▾   SAFETY ▾   INFORMATION ▾   EDUCATION ▾   NEWS ▾   SEARCH ▾   ABOUT ▾

	Go	Past Weather	**dlines**
		Heating/Cooling Days	is the latest instalment of the Watch from Newspaper
Your local forecast office is		Monthly Temperatures	
Greenville-Spartanbr		Records	
		Astronomical Data	

NOWData - NOAA Online Weather Data

1. Location »

View map

- Asheville Area
- Charlotte Area
- Greenville Area
- Pickens Area
- Mount Mitchell Area
- Caesars Head Area
- Bryson City Area
- Ware Shoals Area
- Woodruff Area
- Asheville Reg AP, NC

2. Product »

- ○ Daily data for a month
- ○ Daily almanac
- ◉ Monthly summarized data
- ○ Calendar day summaries
- ○ Daily/monthly normals
- ○ Climatology for a day
- ○ First/last dates
- ○ Temperature graphs
- ○ Accumulation graphs

Variable

Precipitation ⌄

Summary

Sum ⌄

3. Options »

Year range: 2000 - 2021

4. View »

Go

NOWData - NOAA Online Weather Data   [Enlarge results] [Print] ✕

Year	Jan	Feb	Mar	Apr	May	Jun	Jul	Aug	Sep	Oct	Nov	Dec	Annual
2000	3.10	2.33	3.82	5.11	1.27	2.78	2.84	4.45	3.27	0.00	4.25	2.37	35.59
2001	2.63	2.73	5.00	1.32	2.47	2.91	5.50	3.20	4.37	0.60	1.42	2.34	34.49
2002	3.64	1.30	4.36	1.73	3.42	6.13	1.98	2.09	6.05	3.14	4.23	6.40	44.47
2003	1.19	4.47	4.34	5.25	8.36	6.16	10.88	6.80	3.01	2.33	3.89	2.78	59.46
2004	0.83	4.20	2.02	2.95	3.23	7.39	4.68	3.79	13.71	1.11	5.02	3.43	52.36
2005	2.00	2.57	3.33	2.86	1.65	10.09	10.26	5.71	0.34	1.20	3.74	3.51	47.26
2006	3.58	2.55	0.91	4.58	1.69	5.16	2.81	7.12	7.80	2.93	4.52	4.64	48.29
2007	3.35	1.45	4.29	1.77	0.96	2.91	4.85	2.84	3.40	3.02	1.49	4.06	34.39
2008	2.56	3.79	4.51	2.84	1.33	0.85	4.02	5.84	1.70	1.84	1.61	4.74	35.63
2009	2.40	1.87	4.07	3.54	9.18	6.41	2.88	3.69	8.17	5.50	5.26	9.16	62.13
2010	7.00	3.35	4.18	2.24	4.89	1.75	3.54	3.47	4.15	2.94	5.49	1.26	44.26
2011	2.12	2.97	6.95	4.33	2.95	3.83	3.33	3.00	3.74	2.39	5.32	5.11	46.04
2012	3.85	1.59	2.72	4.66	5.82	1.68	5.78	3.39	5.93	4.01	0.85	4.38	44.66
2013	8.58	3.56	3.32	5.88	7.78	8.97	13.69	6.98	3.05	2.19	3.55	7.67	75.22
2014	2.33	3.02	2.30	5.09	3.77	5.39	4.93	3.95	5.87	4.03	3.83	2.40	46.91
2015	3.06	2.78	2.12	4.94	1.35	6.42	2.66	2.77	4.50	7.17	7.82	8.76	54.35
2016	3.29	5.69	1.56	2.50	1.84	2.53	4.39	6.65	0.58	0.52	1.54	2.31	33.40
2017	3.72	0.70	3.92	7.65	7.03	2.71	4.53	6.35	3.75	9.68	1.59	2.47	54.10
2018	4.04	5.57	3.11	4.64	14.68	2.57	6.58	10.41	4.00	5.85	7.16	10.87	79.48
2019	5.28	6.91	2.63	8.97	3.35	6.90	3.69	3.98	0.90	7.78	2.57	4.29	57.25
2020	4.91	7.25	3.39	6.89	5.80	2.40	2.96	9.03	8.27	6.16	3.64	4.01	64.71
2021	3.03	4.45	9.29	1.82	3.37	5.85	5.38	10.95	2.92	5.64	0.88	M	M
Mean	3.48	3.41	3.73	4.16	4.37	4.63	5.10	5.29	4.52	3.64	3.62	4.62	50.21
Max	8.58 2013	7.25 2020	9.29 2021	8.97 2019	14.68 2018	10.09 2005	13.69 2013	10.95 2021	13.71 2004	9.68 2017	7.82 2015	10.87 2018	79.48 2018
Min	0.83 2004	0.70 2017	0.91 2006	1.32 2001	0.96 2007	0.85 2008	1.98 2002	2.09 2002	0.34 2005	0.00 2000	0.85 2012	1.26 2010	33.40 2016

If you have strong computer skills, you can export this data to Excel or any other data-processing software so that you can do your own data analyses. While it isn't absolutely necessary to do this, if you graph the data, you'll understand it better. This data will provide a glimpse into when your rainwater harvesting system will do its most work and when it won't. The data will also give you insight into how much storage capacity you may need to have in order to get through the drier months. In this example, if I lived near Asheville, I would notice that it rains on average more in July and August and less in January and February. I thus would plan for my rainwater harvesting system to collect most of its rain during the summer and the least amount in the winter. Therefore, I would know I should plan to collect more rain during the summer to tide me over through the winter. See the trend?

Next, we'll talk about how much your roof can catch given the amount of rainfall that your location receives. Then we'll compare that to how much water you'll need. First, though, we need to address the times of the year when it freezes.

Freezing Climates

The methods in this book work best during the parts of the year that don't experience freezing temperatures. If you plan to use rainwater as a primary source of water year-round and your climate has deep winters, you'll likely need to involve professionals to create a plan that will either allow you to install your rainwater storage tank in your home (where it will be heated) or have you bury your tank underground underneath the frost line. Another option is to install and power a heater to keep the tank heated during the winter months. If you're unable to employ any of these methods, you'll need to have a secondary source of water during the wintertime (e.g., a well or trucked-in water).

Underground tanks are neither quick nor easy nor inexpensive. Before installing one, consider installing an aboveground system to meet your needs during the nonfreezing times of the year. (You can read more about underground tanks in Chapter 5 to determine if they're better suited for your location/situation as a year-round, long-term solution.) A great resource about the design considerations for underground tanks is *Essential Rainwater Harvesting* by Michelle Avis and Rob Avis. (A book about underground tanks for the QuickRain Blueprint may also be developed soon.)

Climates that freeze for only a few weeks each year can overcome cisterns/pipes freezing by keeping the cistern full to prevent it from freezing solid, using an electronic aerator to continuously move the water so that it doesn't freeze, and/or using dry systems that avoid water sitting in pipes and causing them to burst. More on winterization techniques is covered in Chapter 7: Maintain Your System.

9. UNDERSTAND YOUR CATCHMENT AREA

The size of your catchment area determines how much rain you'll collect regardless of rainfall. If you don't know how big your roof is, follow these steps to use Google Earth to approximate how big your roof is:

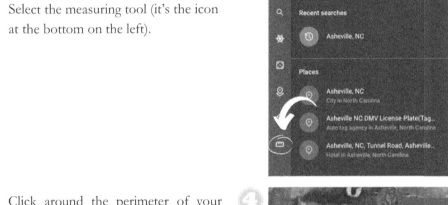

1. Go to Google Earth.

2. Type in your address. Select the 2D view.

3. Select the measuring tool (it's the icon at the bottom on the left).

4. Click around the perimeter of your roof. Change the units to ft². You now have the approximate area of your roof in square feet. Again, I've included some screenshots so that you can see how this looks.

Rule of thumb: 1" of rainfall x 1,000 ft² roof = ~600 gallons of water.

Using your average rainfall data in inches of rainfall and the approximate size of your roof in square feet, use Table 5 to find the *approximate* total amount of rain you can expect to catch. Be careful to note the time period that your average rainfall data falls into. If you have average monthly rainfall data, Table 5 will tell you how much rain you can expect to catch per month. But if you only have rainfall data from a single rain event that you gleaned using a rain gauge, then this table will only tell you how much rain you can expect to catch from a single rain event.

For example, if it rains 4" per month on average in September where I live and my roof is approximately 2,000 ft², I can expect to catch about 4,800 gallons per month if I'm catching rain on all sides of my roof. If it rains 1" today and my roof is approximately 2,000 ft², I can expect to catch about 1,200 gallons from a single rainfall if I'm catching rain on all sides of my roof.

Note that when rain hits your roof, it will follow the downward direction of the roof on the side of the roof that it falls on. If you have a roof with two major slopes, each side of the roof will catch half of the total expected catchment. For example, if I expect to catch 1,200 gallons during a single rainfall, I'll catch roughly 600 gallons on one side of my roof and 600 gallons on the other side. I can place a tank on either side of my roof because I know that rain will end up in tanks on both sides.

Rain tanks capturing rain runoff from both sides of a roof.

The actual amount you catch will almost always be less than your expected amount due to inefficiencies in the roof (asphalt roofs are more porous than metal roofs, for example), not all of the rain making it to your gutters, and other factors. Regardless, having an idea of how much total rainwater you can expect to catch can help you understand the potential of your roof and how you can modulate and change your rainwater harvesting system at a later date.

TABLE 5. EXPECTED AMOUNT OF RAINWATER THAT CAN BE CAUGHT

Amount of rainfall in inches	Area of roof (or any catchment area) in square feet			
	1,000 ft^2	1,500 ft^2	2,000 ft^2	2,500 ft^2
0.5"	300 gallons	450 gallons	600 gallons	750 gallons
1"	600 gallons	900 gallons	1,200 gallons	1,500 gallons
1.5"	900 gallons	1,350 gallons	1,800 gallons	2,250 gallons
2"	1,200 gallons	1,800 gallons	2,400 gallons	3,000 gallons
2.5"	1,500 gallons	2,250 gallons	3,000 gallons	3,750 gallons
3"	1,800 gallons	2,700 gallons	3,600 gallons	4,500 gallons
3.5"	2,100 gallons	3,150 gallons	4,200 gallons	5,250 gallons
4"	2,400 gallons	3,600 gallons	4,800 gallons	6,000 gallons
4.5"	2,700 gallons	4,050 gallons	5,400 gallons	6,750 gallons
5"	3,000 gallons	4,500 gallons	6,000 gallons	7,500 gallons

10. UNDERSTAND YOUR WATER NEEDS

Now that you understand how much rainwater you can expect to catch, use the following tables to gain a basic understanding of how much water average Americans use per week, per month, and per year for their garden and inside their home. Refer to Table 6 if you're using rainwater to water a garden/yard and Table 7 if you're using rainwater inside the home. Then compare the amount of rain you expect to catch with the numbers in Table 5. For example, I just learned that I could expect to catch about 4,800 gallons per month in September from my 2,000 ft² roof.

Using Table 7, I find that a family of four will need about 3,000 gallons of water for indoor use per month. Rainwater in my location will likely meet all of my indoor needs, and the wiggle room could account for variations in rainfall, catchment efficiency, and other risks. I feel confident about September.

Consider creating your own table of your expected water catchment per month based on your location's rainfall and your catchment area and then see if it will meet your family's needs for that month.

TABLE 6. BALLPARK GARDEN WATERING NEEDS (ASSUMING DURING THE SUMMER MONTHS) [3]

Garden/yard size	Weekly use	Monthly use
Small garden (~100 ft²)	60 gallons	250 gallons
Medium vegetable garden (may have livestock) (~500 ft²)	300 gallons	1,250 gallons
Large garden (may have livestock) (~800 ft²)	500 gallons	2,000 gallons

TABLE 7. BALLPARK INDOOR USE WATER NEEDS

(based on figures from *Create an Oasis with Greywater* by Art Ludwig; numbers have been rounded up for ease of use)

Application	Frequency used (per week)	Volume per use	Weekly use per person	Monthly use per person	Yearly use per person
Top-loading washing machine	1.5	30 gallons	45 gallons	180 gallons	2,160 gallons
Shower	5	13 gallons	65 gallons	260 gallons	3,120 gallons
Bathroom sink	21	0.5 gallons	11 gallons	45 gallons	500 gallons
Toilet	35	1.6 gallons	35 gallons	250 gallons	2,600 gallons
Drinking & cooking	7	1.5 gallons	11 gallons	45 gallons	500 gallons
Total per person			180 gallons	750 gallons	9,000 gallons
Total for 2 people			360 gallons	1,500 gallons	18,000 gallons
Total for family of 4			750 gallons	3,000 gallons	36,000 gallons

- How to Lower Water Needs and Consumption

If you're using rainwater to water plants or irrigate your yard or property, it's best to grow plants that are native to your location—those plants are accustomed to the amount of rainfall where you live. You can increase the ability of your soil to hold more water by keeping it covered with living plants. In effect, you'll be building living sponges! To support healthy mini-ecosystems, maximize the biodiversity of plants on your property. Even if you plan to use just a rain barrel for harvesting, understanding the interplay between your garden's water needs, your catchment potential, and your localized amount of rainfall will help you better understand your rainwater harvesting system and how it will interact with your ecosystem.

If you're using rainwater for indoor use and if you won't be able to capture the amount of water your family needs, you'll need to support your rainwater catchment with a secondary source of water (which is actually always a good idea in general) or lower your family's water usage. Some ideas for how to do this include using low-water-volume fixtures and appliances (e.g., low-water-volume shower heads, washing machines, dishwashers, etc.), taking shorter showers, and not leaving faucets running.

ACTION GUIDE FOR STEP 4: LEARN THE BASICS OF SAFETY AND USE

Step 1. Double-check your basement and foundation for any leaks or cracks. If soil is pulling away from your foundation, fill in the cracks with clay. Fill any cracks in the basement by yourself or call a waterproofing professional. Make sure that your foundation is good to go before playing with water diversion.

Step 2. The next time it rains, observe where the rain flows away from your downspouts and watch how it flows, puddles, and pools in your yard and property. Figure out if those puddles or pools could be redirected somewhere more useful—recognize any garden beds, trees, or troughs where rain could be used productively. Identify the best downspout on your home where a rain storage tank could go.

Step 3. Find the average monthly rainfall from the past 20 years for your location using the steps outlined in the Understand Your Climate section. Purchase a rain gauge from the hardware store if you don't have one already.

Step 4. Calculate your approximate roof size if you don't know it already. Estimate how much rain you can expect to catch per month using Table 5. Begin to estimate how big of a tank you may need depending on the average rainfall in your location. (Final estimates based on budget and space will be determined in the next chapter.)

Step 5. Determine if the amount of rain you can expect to catch will meet your family's water needs for indoor use and/or gardening. If you'll be using rainwater for both outdoor and indoor use, add up the amount of water needed per week and per month from Tables 6 and 7. Estimate if the amount of rainwater you expect to catch will provide enough wiggle room based on weekly and monthly usage. Determine a secondary source of water if needed (e.g., arrange for trucked-in water).

SCALE AND SOURCE YOUR SYSTEM

CHAPTER 5

N ow it's time to move on from just learning. Let's get to building! First, you'll need to source or purchase the materials and equipment for your system. See Table 8 to determine which components you'll need to buy depending on your end-use goals. The methods described in this book assume that you'll use your current roof and gutter system to harvest rainwater and so does not consider those components to be part of the sourcing process. That said, your particular situation may require sourcing a new roof or gutters or upgrading your gutters if you would like to have seamless gutters.

Unless there's a hardware store near you that stocks rainwater harvesting equipment (or a specialized rainwater harvesting store), you'll likely purchase most of your specialized equipment such as filters and tanks online. This chapter will provide examples of what to search for online and examples of online stores where components can be found and bought based on your budgetary considerations.

The contents in this chapter and in Chapter 6: Install Your System provide greater context as to where different components go and how different components plug and play together. Understanding that context can help you in your sourcing decisions. Chapter 6 includes installation guides as well as different scenarios of where different types of filtration systems and tanks and foundations are used. It's a good idea to read these two chapters with your imagination and paper and pencil at hand! Once you have an idea of which components go where and which components you'd like to have, sketch out some ideas and start designing your dream rainwater harvesting system.

TABLE 8. COMPONENTS YOU'LL NEED TO PURCHASE DEPENDING ON YOUR END-USE GOALS

End-Use Goals	Components You Will Need to Purchase
Outdoor Non-Potable Use	1. Screens and filters for gutters, downspouts, and tanks 2. Storage tank 3. Storage tank foundation 4. Pipes or hoses (inlet, overflow, outlet) 5. First-flush system (optional) 6. Pump (optional)
Indoor Non-Potable Use	1. Screens and filters for gutters, downspouts, and tanks 2. Storage tank 3. Storage tank foundation 4. Pipes or hoses (inlet, overflow, outlet) 5. First-flush system (optional) 6. Pump (less optional) 7. Indoor-use finer filtration (around 20 – 100-micron filters) 8. Indoor-use pipes
Indoor Potable Use	1. Screens and filters for gutters, downspouts, and tanks 2. Storage tank 3. Storage tank foundation 4. Pipes or hoses (inlet, overflow, outlet) 5. First-flush system (optional) 6. Pump (even less optional) 7. Indoor-use even finer filtration (around 5 – 20-micron filters) 8. Indoor-use pipes 9. Potable water treatment

CONSIDERATIONS ABOUT AND WHERE TO PURCHASE EACH COMPONENT

1. SCREENS AND FILTERS FOR GUTTERS, DOWNSPOUTS, AND TANKS

Screens for your downspouts and tanks/cisterns are essential for filtering rainwater for leaves and other big debris (branches, twigs, etc.) before the water enters the tank.

Gutter Screens

Gutter screens and gutter guards are helpful for keeping leaves and other organic matter out of gutters. While there's a debate about how useful and safe these screens are, if you live in an area that's prone to large rain events or in an area with lots of foliage, in the long run, screens can save you time and energy from having to continually clean out your gutters. That said, gutter guards and screens can keep organic matter close to the trough of your roof, and that may lead to issues with your roof if the organic matter rots before it's removed. However, if you wish to have higher quality rainwater, you should be cleaning out your gutters at least twice a year anyway, and if you have more foliage, you'll need to clean them out as often as once a month in the rainy season. (I live underneath a large wisteria that provides lovely shade in the summer, but its leaves fill my gutters at an alarming rate.)

Rule of thumb: Don't install gutter guards or gutter screens unless your climate or foliage necessitates it.

If you believe gutter guards or screens will save you time and energy in the long run, consider installing them in tandem with your rainwater harvesting system. While they're not foolproof—they only reduce rather than prevent debris and leaves from falling into your gutters—gutter guards or screens may allow you to harvest higher volumes of higher-quality rainwater over the years. Full gutter screens will cost an average of $1,000, with the price depending on the quality of the materials.

You can install some gutter guards yourself. Typically, this is possible if your home is single-story or your roof isn't very steep. You can purchase gutter screens by the yard at hardware stores like Home Depot and then follow the manufacturer's instructions for installing them.

Downspout Filters

Downspout filters add another layer of filtration as rainwater travels from the gutters to your tank or cistern. Most of these filters fit in line with your downspout. That is, you would cut your downspout and fit in a filter so that as rainwater flows down your downspout, it passes through the filter. Downspout filters come in various sizes, so measure the diameter of your downspout before making your purchase. Cheaper, shorter-lasting filters cost about $6 – $10, whereas more expensive, longer-lasting filters cost about $50 – $80. Search for "downspout filter," "rainhead gutter," or the brand name Leaf Eater at your hardware store or online.

A version of a rain head (this one made of stainless steel), which can act as a downspout filter.

Once the water flows off your roof and into your gutters and downspouts, it will pass into your tank or cistern. Some rain barrels come with a screen already installed in them, but if yours doesn't have one, you can buy filters that you can insert into the inlet. Look for "basket strainer for rain tanks" or "rain barrel screen" at your local hardware store or online. The holes in the filter should be less than 1/16" (less than 1.5 mm) to prevent mosquitos from entering. They will cost around $10 – $20, which is an inexpensive yet effective method to filter rainwater just before it enters the tank.

A screen on a rain tank filtering out leaves and debris before rainwater enters the tank.

Use a combination of a downspout filter and a screen on your tank to achieve the best balance between safety, maintenance, cost, and the ability to filter rainwater before it enters the tank. Add gutter guards or screens if you find yourself cleaning out your gutters from leaves and debris too often (i.e., more than once every three months). Even then, clean out your gutters at least twice a year and keep foliage and tree branches off your roof and gutters by cutting back limbs and branches.

Materials Used to Make Gutters and Downspouts

Aluminum gutters are the most common type of gutters. Your gutters might also be made of zinc, galvanized steel, vinyl/PVC, or copper. Your gutters will be generally safe and suitable for collecting water for non-potable and potable uses if they're made of aluminum, zinc, or galvanized steel. If you have vinyl/PVC gutters, they may have traces of lead in them from the manufacturing process. If you have copper gutters, unless you know for sure that they're coated, be aware that copper can leach into the rainwater and that too much copper in water can lead to heavy metal poisoning. Because of this, don't water edible plants with rainwater or drink rainwater if you have vinyl/PVC gutters or non-coated copper gutters. If you don't know what your gutters are made of, it's better to err on the side of caution and use your harvested rainwater only for non-potable uses such as irrigating nonedible plants and flushing the toilet until the gutters are replaced. (Note that the average cost of replacing gutters is $1,000.) Start harvesting rain with the gutters you have now, as they are likely suitable for plenty of non-potable uses already. They might even be suitable for potable uses.

2. STORAGE TANKS

Rainwater Storage Tanks Overview

A tank or cistern is the container that stores the rainwater after it flows from the roof, into the gutters, and down the downspout. When you hear the phrase "scaling the system," "scaling" means determining the size of your storage tank. The instructions in this book apply to any size rain barrel, whether it holds 55 or 5,000 gallons of water.

Tanks come in all shapes and sizes. Tanks should be opaque to avoid algae and bacteria growth and should be tightly covered to prevent mosquitos, animals, and children from entering the tank. (Note that tanks should still allow for a way for air to pass in and out of the tank.) Choose your barrel/tank/cistern's size and material in accordance with why you're storing rainwater and how much you want to store. For example, if you simply want to water a small garden (~100 ft²), opt for a 55-gallon plastic drum that will last 1 – 2 years. If you want to water a backyard-sized vegetable plot or backyard animals, collecting rain in a 275-gallon IBC (intermediate bulk container) tote or tank is a good option. If you plan to collect rain to water a large garden for several years, consider a larger, longer-lasting tank made of polyethylene or metal that holds 500 – 3,000 gallons and can last more than 10 years. For most indoor uses, consider larger tanks that hold 1,000 – 5,000 gallons to account for all of your family's water needs.

While differing watering needs and climate considerations make it tricky to quickly estimate tank sizes for a home or homestead, know that rain tanks used to water medium-sized, backyard gardens (200 – 500 ft²) are generally 300 – 500 gallons, rain tanks used to water large (>600 ft²) gardens can be as large as 500 – 1,000 gallons, and rain tanks for indoor use can be as large as 1,000 – 5,000 gallons (or more). You can also start with a smaller tank and add more storage capacity later.

If you've never collected rainwater before, start with a rain barrel to learn how to install and maintain a basic rainwater harvesting system. If you already have some experience and wish to dive into bigger projects, feel free to start with a bigger tank. For rainwater to be a primary water source, opt for a larger tank rather than a smaller one, as you will need more storage to overcome droughts or dry spells. Determining the exact tank size you need will require trying several sizes first; it's very common to end up with several tanks.

Typical residential rain tanks range from 55 gallons to around 3,000 gallons and can cost from $0 to $6,500 depending on the quality of the tank. As an estimate for those starting out, plan to spend $300 – $500 for just the tank(s) for a medium-sized garden (200 – 500 ft^2). Those already planning for upgrades or who want to use rainwater for indoor water needs can plan to spend $800 – $6,500 depending on the size of the tank and the material. The more volume of storage per tank, the lower the cost of storage per gallon.

Rain Barrels ($0 – $100)

Rain barrels (40 – 80 gallons per barrel): Purchase a prefabricated rain barrel that already comes with screens and overflow valves either online (e.g., Amazon, Wayfair) or from department stores like Target or Wal-Mart (in the garden section) or hardware stores like Home Depot or Lowe's (in the garden or plumbing section). Purchasing a prefabricated rain barrel saves you the time of finding a barrel and installing inlets, outlets, screens, and valves yourself. These barrels cost about $100 each.

Alternatively, you can save money by fabricating your own rain barrel. Plastic trash cans with a lid work well, as do plastic drums. Recycled 55-gallon drums can be sourced for free from bottling plants or car washes, or you can find new or recycled barrels for a reduced cost at vineyards or distilleries. You can also check Craigslist or Facebook Marketplace. Cities often have rain barrel programs, offering free rain barrels and classes on how to install and use them; check online to see if your city or county offers these free barrels. Barrels should come with a lid to allow access for cleaning but be sealed tightly enough to prevent mosquitos, animals, and children from getting in. Before purchasing a recycled barrel, double-check its past life, as plastic (and even metal and ceramic) will contain traces of what it used to store. Opt for food-grade barrels. If in doubt about a barrel's past life, buy a different barrel or buy a new barrel.

A version of a prefabricated rain barrel with inlet screens already installed.

Rain barrels are an excellent choice for beginners, those wanting to supplement their municipal or well water needs with rainwater, or those just needing to water a small (~100 ft^2) garden plot. They are quick to set up—they take about 30 – 120 minutes to install, making for a great Saturday afternoon project—and are an inexpensive option for harvesting rain right away. However, during large rain events, they can overflow quickly. If you've already set up your rain barrel and want to capture more rain than it catches currently, it could be time to install more rain barrels or a bigger tank.

IBC totes ($175 – $350)

IBC totes (275 – 330 gallons per tote): If you're looking to install tanks in the 275 – 500-gallon range, a great option is what's called an "intermediate bulk container" or IBC "tote" or IBC "tank." IBC totes come in many shapes and sizes, but they have one thing in common: they can be lifted and stacked with a pallet jack quickly and easily. They're used industrially for transporting large volumes of liquid, such as milk, pharmaceuticals, and pesticides. They're particularly useful for rainwater harvesters because IBCs are relatively inexpensive, come with lids and a spigot at the bottom already installed, and can hold up to 330 gallons per tank (it's hard to find IBCs that hold less than 275 gallons per tank). However, it should be noted that IBCs need to be somewhat reconfigured before they're ready for use as a rainwater tank. That's because of two reasons: IBCs are not automatically sealed from insects/animals and a vent will need to be installed. If your IBC tote is translucent, you'll need to do a third thing, namely paint it or cover it to avoid algae growth. Configuration ideas are explained in Chapter 6.

A translucent IBC tote capturing rainwater off a garden shed.

Consumers like you and I can buy IBCs from recyclers and salvage yards; advertisements for them are common on Craigslist and Facebook Marketplace (just search for "IBC tote" or "IBC tank"). They can also be bought online new—search for "buy IBC tote," and several online vendors will pop up. A 275-gallon tote will cost around $175 used or $350 new. Because of the variety of liquids they typically carry, double-check an IBC tote's past life by asking your vendor what was previously stored in the tank. If they don't know what was in it, skip it. Opt only for food-grade totes even if you're only using them for non-potable water. Just like with rain barrels, if you're in doubt about a tote's history, buy a different one or a new one.

If you have a medium-sized vegetable garden or a backyard homestead, placing IBC totes underneath your downspouts is a great option. However, many IBCs are white and translucent. Since tanks should be opaque to avoid algae growth, try to buy black IBC totes or paint your white IBC totes any color until they're opaque (common spray paint for plastic works very well). Another option is to wrap the tote in plastic or fabric like you would a present. Standard 275-gallon IBC totes have an international standard size of roughly 40" L x 48" W x 46" H. Compared to other tank sizes and shapes, IBC totes fit nicely into the bed of a truck or the back of a van, giving you the flexibility of transporting the tank yourself from the vendor to your home. Other types of rain tanks often need to be delivered because of their shape and/or size.

Cisterns
(Medium Budgets up to $4,500; High Budgets up to $20,000+)
Rain tanks can be considered a do-it-yourself size up to ~2,500 gallons. But while you can still buy and order larger tanks yourself, tanks larger than ~2,500 gallons will need strong foundations that may need professional-level installation.

Polyethylene tanks
(medium budgets: 100 – 1,000 gallons per tank; high budgets: 1,000 – 20,000 gallons per tank): These tanks are longer-lasting—they can last for 15 to 20 years—yet are also a good economic solution for storing rainwater. High-density polyethylene (HDPE) is the most common PE (polyethylene) material used to build water tanks, as it is strong and impact-resistant enough for most residential and backyard rainwater harvesting needs. Most rain tanks made of HDPE can be food-grade and approved for potable water storage by the National Science Foundation under the code "NSF 61". Although they can degrade in direct sunlight, HDPE tanks can be treated to avoid UV degradation. And while they are best placed in shady locations, they can also be encased in wood or masonry to further protect against UV rays. HDPE tanks are also relatively light and can be moved into place just by rolling or pushing them along the ground.

Compared to other types of tank materials such as galvanized steel or concrete, tanks made of PE are lighter and therefore cheaper to deliver (and easier to transport and install). You may find that most tanks made of PE are round and available in a variety of diameters and heights. Factor this into where you think your tank(s) will be placed in case fitting round objects next to square buildings could pose a problem for you. If a round shape would be an impediment, a slimline tank might be a great option for you. Slimline tanks come in rectangular shapes and are usually slimmer than they are tall, fitting nicely next to homes and attaching directly to gutters more easily. Just be wary that slimline tanks have a higher center of gravity and therefore are more prone to tipping over. This can be mitigated by constructing a cage around the tank or fastening it to a building with a tie to keep it upright.

Tanks made of PE come in a variety of colors and shapes.

A cylindrical polyethylene tank held in place with a tie.

Opt for one that's already opaque to avoid having to paint it opaque. It's true that PE tanks are not the most beautiful tanks out there, especially compared to the glimmer of a steel tank. However, PE tanks won't rust over time, unlike galvanized steel (i.e., zinc-coated steel), which can be susceptible to rust (over about 15 years). But PE tanks are not wildfire-resistant and will likely melt if there's a wildfire, so if your location *is* prone to such fires, tanks made of metal (or other materials, such as fiberglass or concrete) are better options. Also consider a PE tank's moderate 15 – 20-year lifespan and its contribution to plastic waste.

If you decide to purchase a PE tank, you'll likely buy one online (you can search for "rainwater tanks for sale" and browse the options for plastic rainwater harvesting tanks from several online dealers, such as Plastic-Mart.com or The Tank Depot), through a hardware store (such as Lowe's or Tractor

Supply Co.), or through a professional rainwater harvesting company (search for "rainwater harvesting professionals near me" on your search engine if you live in a major metropolitan area; there should at least be a few). Browse their prices—you may find that online retailers have cheaper prices than brick-and-mortar stores or vice versa. Medium-range polyethylene tanks run from around $400 for a 100-gallon tank to $1,200 for a 1,000-gallon tank. Once you've bought your tank online or from a hardware store, you'll likely need to have it delivered to your home and pay for delivery/shipping. Extra points if you can get it to your backyard or homestead by yourself! Tanks made of PE simply go up in cost as they go up in size, so the larger the tank is, the lower the cost per gallon of storage will be. At around 1,500 gallons, a PE tank costs about $1,500. At 2,500 gallons, a PE tank costs about $1,800. Tanks made of PE go up to around 20,000 gallons per tank; that size costs around $36,000. PE tanks last for about 15 years, so if you decide to spend what a large tank costs, also consider metal tanks since they can last up to 30 years.

Metal tanks
(medium budgets: 80 – 150 gallons per tank; high budgets: 150 – 100,000+ gallons per tank):
Metal rain tanks come in a variety of metal types, usually as galvanized steel (which is zinc-coated steel that's prone to rust once it hits 10 – 15 years) or stainless steel (which is very unlikely to rust at all). Not only are they easy to find and as such are an economical choice for rain tanks, metal tanks can last for 20 – 30 years. They aren't prone to UV degradation like PE tanks are, plus metal tanks can be beautiful additions to a residential property or homestead (unlike PE tanks, which are often considered ugly and therefore are hidden in a shady spot in the backyard.) If the location of the tank and its aesthetic value for neighbors are a concern, metal tanks are a great option.

Cylindrical metal tanks capturing rainwater on a commercial property.

Slimline tanks (which can be made of either polyethylene or metal) save space when placed next to buildings

Metal tanks come in slimline varieties, too, which look nice and perform well next to buildings.

To use metal tanks for potable water applications, you'll need to find a metal tank that already comes with a potable water liner or coating on the inside—look for liners that are marked "NSF 61 approved." Many metals tanks come ready for potable water storage; the metal usually has a low impact on how the water tastes.

If you're choosing a metal tank, give special consideration to the materials used in the piping! The wrong kind of metal pipes can cause a metal tank to corrode. A solution for this is to use plastic piping for the last few yards leading up to the tank. The most common example of this potential problem is aluminum downspouts; aluminum in contact with steel can cause the steel to corrode. Rather than having them lead directly into the tank, attach a flexible plastic pipe to the aluminum downspouts and then attach the plastic pipe to the steel tank.[2]

If your location is prone to wildfires, metal tanks are a better option than PE tanks due to the fire resistance offered by metal tanks. When looking for wildfire-protected tanks, look for tanks rated "NFPA-22." Specific manufacturers to look for are Aquamate and Pioneer Water Tanks, though any tank rated NFPA-22 is suitable. These tanks don't need to be only used for wildfire-prone areas, either; also consider these manufacturers for a rain tank for domestic use.

You'll achieve around 150 gallons of rainwater storage with a metal tank that costs up to about $1,200. These tanks are not the most economical solution for smaller volumes of rainwater storage, whereas

PE tanks and IBC totes are. Metal tanks are better suited for rainwater storage in the ~500 – 5,000+-gallon range, and again, they also last 10 – 20 years longer than PE tanks or IBC totes do. If you want longer-lasting, more permanent rainwater storage and use, metal tanks are a solid option.

You can find metal tanks online from rainwater tank distributors such as Plastic-Mart.com or The Tank Depot (search for "rainwater tank steel" or similar terms), at hardware stores such as Lowe's or Tractor Supply Co., or through a professional rainwater harvesting company. Most places that carry PE tanks also carry galvanized steel tanks; some carry stainless-steel tanks.

If you decide to purchase one of these tanks, you'll likely need to have it delivered to your home—metal tanks must be handled carefully as they can dent easily during shipping or installing. Remember that even though galvanized steel is coated with zinc to prevent rusting, it can still rust over time (i.e., 10 – 15 years in high humidity). Because stainless steel has properties that render it very inert, it resists rusting for much longer—100 years or more—but it costs twice as much. Metal tanks are manufactured and distributed in certain locations (Texas is a common place for steel tanks), so if you live outside these locations, you may pay higher shipping and handling costs.

Metal tanks made of galvanized steel and stainless steel are excellent options for tanks that range from holding hundreds of gallons to thousands of gallons. While they're more expensive than PE tanks (especially when storing ~100 – ~300 gallons of rainwater), their lifespan makes them excellent options when storing rainwater in the ~500 – ~3,000-gallon range. Galvanized steel rainwater tanks for 500 gallons go for around $2,000 (stainless steel is about twice that); tanks holding 3,750 gallons sell for around $9,000. Once steel tanks enter the ~5,000-gallon range, prices differ due to manufacturing costs, shipping costs, and availability, but plan to spend about $10,000. For residential- and homestead-scale systems, multiple tanks can be installed over the course of several years. Try starting with tanks in the ~1,000-gallon range to keep startup costs low and then add tanks over time as storage needs increase. Metal tanks can be very large, going up to 100,000 gallons or more.

Properties looking to fight wildfires typically start with tanks in the 5,000-gallon range. With larger tanks, you can use them for the home and garden while storing enough reserve water for fighting fires. Many metal tanks of this size come in configurations that allow them to be installed on-site. These tanks aren't as bulky or as prone to damage during shipping. They're called "bolted steel" tanks and are assembled on-site as opposed to arriving on-site already assembled. If this kind of tank appeals to you, look to purchase one on Plastic-Mart.com, Pioneer Tanks's website, or Aquamate's website.

An example of a bolted steel rain tank. This type of tank is often assembled on site.

Underground Tanks

You have may noticed that some rainwater harvesters install their tanks underground and have wondered if this solution is for you. Underground placement has certain advantages that can make the complex installation process worth it. If you live in a location where accessing rainwater during freezing winters will be a problem, for example, then burying a tank underneath the frost level will help keep rainwater at a useable temperature year-round. Consider doing this if rainwater is your primary source of water or a deeply necessary secondary source of water. If you plan to collect large amounts of rainwater (e.g., your tank will hold 5,000 or more gallons) and you have limited space on your property, storing tanks underground can allow you to save space.

It's generally not a good idea to install underground tanks yourself—this kind of installation requires the expertise of professional rainwater system installers. If cost and speed are primary concerns, keep your rain tank above the ground for now. That said, if underground tanks could indeed be your solution, you can read more about underground tanks in *Essential Rainwater Harvesting* by Michelle Avis and Rob Avis. Search for rainwater harvesting professionals near you and ask if they have experience in installing underground tanks. They will guide you through the process of surveying your property, sourcing the right materials, installing the underground tank, and installing the right conveyance system and pump to meet your needs.

Underground tanks in this book are considered food for thought. While they work well for some, they're a solution that requires plenty of time, money, and energy. For most residential and backyard applications, it can be best to install tanks aboveground to avoid the complexity and cost of installing a tank underground. Consider an underground tank only if you find that the cost and complexity are justified.

Other Tank Materials

Besides plastic and metal, rain tanks around the world are made of fiberglass and concrete. However, if your rainwater harvesting system is intended to lower your water bills or capture rain for plants and animals, tanks made of plastic or metal are typically the most cost-effective options you can install quickly on a do-it-yourself scale. Plastic and metal tanks are also the most common and therefore the easiest tank materials to find.

Concrete tanks are typically precast by a manufacturer and then delivered and installed with a crane. They're immediately useful for potable water applications—they even add minerals to the water for improved taste as well as health benefits. They can also last for decades.

Rain tanks made of concrete can last for decades but can be harder to find.

If you're looking for a concrete tank, you'll likely need to find a manufacturer near you (search for "rainwater harvesting tank concrete near me") and call for a quote. Ask the manufacturer what types of cement they use for their concrete and be wary of toxic cement materials such as fly ash, which can contain heavy metals and toxic metals such as arsenic and lead.

Fiberglass tanks are also typically precast by a manufacturer and can be custom-made to your specifications, then delivered and installed with a crane. They're also immediately useful for potable water and can last for decades. They are also among the most expensive tank materials precisely because of their desirable qualities. Similar to what you would do for a concrete tank, search for "rainwater harvesting tank fiberglass near me" and call for a quote.

More thorough information on tank materials and their effects on water quality can be found in *Essential Rainwater Harvesting* by Avis and Avis.

How Big Should My Tank Be?

The short answer is "As big as your budget allows."[1] For many residential and homestead applications, having more water is better than having less water, and storing rain when it rains ensures you'll have access to it when it *doesn't* rain. That said, if your budget and space are tight, consider these guidelines:

A rain barrel or two is useful for small plots and gardens (~100 – 200 ft²). IBC totes—up to four or so—are useful for medium-sized gardens and animals (up to ~500 ft², with some backyard livestock, like chickens or goats). This is also a common storage solution for homes with one or two people looking to use rainwater inside the home. However, because IBC totes are restricted in size, keeping multiple totes clean and maintained can be troublesome. Larger cisterns may be needed.

Cisterns of around ~500 – ~5,000 gallons are useful for larger gardens and homesteads (~800 ft²). Homes with two adults and two children who grow their own food and raise their own chickens and goats may find that 5,000 gallons offers enough water storage. Your family may do fine with less, or you may need more.

Conditions such as your climate, water needs, efficiency of water use, and efficiency of water catchment can increase or decrease your water storage needs quite significantly. You'll need to determine the most accurate answer for your situation. Look at the tables below for general guidelines that will help you find the sweet spot for your rainwater harvesting tank(s). Note that these numbers are conservative estimates! Your personal water needs, along with the climate of your location, may dictate higher or lower water storage needs. The more arid your climate is, the bigger your tank will need to be, because you'll need to store more rain for drier times of the year. If your budget and space allow for it, opt for a bigger tank than what's in the table so that you can store as much water as you possibly can. Also, if you depend on rain as your primary source of water, always err on the side of more storage than less, especially if there are long stretches of dry months where you live.

Optimizing rainwater catchment to climate and use to water needs usually involves trying different sizes of tanks over time. Many rainwater harvesters, homesteaders, and gardeners do not do extremely fine calculations to optimize how to best harvest and use rainwater even when it's their primary source of water. However, you can better optimize by calculating the exact size of the tank you'll need based on granular data regarding your local climate, rainfall intensity, catchment efficiency, and water use (among other factors). Both *Essential Rainwater Harvesting* by Michelle Avis and Rob Avis and *Rainwater Harvesting for Drylands and Beyond, Vol. 1* by Brad Lancaster provide advice on how to do these calculations.

If rain will be your primary source of water, also consider having a secondary source of water (such as a well, stream, or municipal source) or a backup strategy (such as having water trucked in) so that you'll have peace of mind as you gradually increase the size and scale of your rainwater harvesting system (and continue to learn the rainfall patterns of your location).

TABLE 9. BEST TYPE OF TANK BASED ON BUDGET

Budget	Rain Barrel	IBC Tote	PE Tank	Metal Tank
Low	$0 – $100 for 40 – 80 gallons			
Medium		$175 – $350 for 275 – 330 gallons per tote	$300 – $1,200 for 100 – 1,000 gallons per tank	$300 – $1,200 for 80 – 150 gallons per tank
High			$4,500 – $36,000+ for 4,000 – 20,000 gallons	$4,500 – $100,000+ for 1,600 – 100,000+ gallons

TABLE 10. FEATURES OF DIFFERENT RAINWATER STORAGE TANKS

Feature	Rain Barrel	IBC Tote	PE Tank	Metal Tank
Recommended usable lifespan	1 – 5 years	2 – 5 years	15 – 20 years	20 – 30 years (more for stainless steel)
Suitable for potable water	Not often	Sometimes (look for NSF 61 rating and food-grade rating)	Yes; look for NSF 61 rating	Yes; look for NSF 61 rating
Prone to rust	No	No	No	Sometimes; galvanized steel can rust in high humidity after ~10 – 15 years
Prone to UV degradation	Not within usable lifespan	Yes; although totes can be treated for UV protection, it's still best to position them in shady areas	Yes; although PE tanks can be treated for UV protection, it's still best to position them in shady areas	No

	Online (Amazon, Wayfair), hardware stores (Home Depot, Lowe's, Tractor Supply Co.), secondhand recycling centers (soda bottling plants, vineyards, etc.)	Online (for new, search for "buy IBC tote new") or via salvage companies (for new or used, search "buy IBC tote" on Craigslist or Facebook Marketplace)	Typically bought online (common online vendors are Plastic-Mart.com, The Tank Depot, Rain Harvest Systems), at some hardware stores like Lowe's or Tractor Supply Co., or via your local professional rainwater harvesting company (search for "rainwater harvesting professional near me")	Typically bought online (common online vendors are Plastic-Mart.com, The Tank Depot, Rain Harvest Systems, or Pioneer Tanks or Aquamate), at some hardware stores like Lowe's or Tractor Supply Co., or via your local professional rainwater harvesting company (search for "rainwater harvesting professional near me")
Where to buy				
Useful/cost-effective to store 100 gallons	✓	X	X	X
Useful/cost-effective to store 500 gallons	X	✓	✓	✓
Useful/cost-effective to store 1,000 gallons	X	X	✓	✓
Useful/cost-effective to store 5,000 gallons	X	X	✓	✓

From having read Chapter 4, you should have determined about how much rainwater you can expect to catch (based on your rainfall and roof size) as well as estimated how much water your family may need. For example, in my location with my roof size, I can expect to catch 4,800 gallons of rainwater for the month of September, and I know I may need 3,000 gallons of water per month for my family of four. In addition to those calculations, now use Table 11 to determine the size of tank(s) you may need depending on your application(s) and how much space you may need for a tank or tanks of that size. If you wish to collect rainwater for both indoor and outdoor uses, add up how much rain you'll need given the size of your garden *and* the size of your household. For example, if I had a large garden and a four-person family, I would aim for—at a minimum!—5,000 gallons of rainwater storage in case of dry spells or droughts (2,500 gallons + 2,500 gallons = 5,000 gallons). My example for my situation would work because my family needs 3,000 gallons and I can expect to catch 4,800 gallons in a month. That also allows for some wiggle room in case my catchment efficiency is not as high as I had predicted it would be or it doesn't rain as much as it usually does.

TABLE 11. NUMBER AND SIZE OF TANKS NEEDED DEPENDING ON APPLICATION

Application	Rain Barrel	IBC Tote	PE Tank	Metal Tank	Recommended Area Needed for Rain Tank
Small garden (~100 ft²)	(1) or (2) 40 – 80-gallon barrels	(1) 275-gallon tote	Not cost-effective	Not cost-effective	~10 ft²
Medium vegetable garden (may have backyard livestock) (~500 ft²)	Useful for beginners; otherwise, not recommended	(2) 275-gallon totes	(1) 500-gallon tank (cost-effective at ~500 gallons)	(1) 500-gallon tank (cost-effective at ~750 gallons)	~12 – 15 ft²
Large garden (may have livestock) (~800 ft²)	Useful for beginners; otherwise, not recommended	Up to (5) 275-gallon totes	(1) ~2,500-gallon tank	(1) ~2,500-gallon tank	~30 – 40 ft²

| 1 – 2 people wanting indoor use (assuming 180 gallons/person/wk) | Not very useful for indoor water volume needs | Up to (5) 275-gallon totes | (1) ~1,000 gallon tank | (1) ~1,000 gallon tank | ~20 – 25 ft² |
| 2 – 4 people wanting indoor use (assuming 180 gallons/person/wk) | Not very useful for indoor water volume needs | Up to (9) 275-gallon totes | (1) ~2,500-gallon tank(s) | (1) ~2,500-gallon tank(s) | ~30 – 40 ft² |

3. STORAGE TANK FOUNDATIONS

Water is heavy, and when rain tanks are filled with water but not properly stable, they can exert forces on the ground in ways that can be damaging to your home or yard (and the rain storage tank you invested in!). Having a proper weight-distributing foundation for the cistern prevents this. Luckily for us, proper foundations can be installed on a DIY basis. Before installing your foundation, find a spot beneath a downspout on your building that is flat, level, and clean. Remove any large or medium rocks that may puncture the bottom of your tank. If the ground is not level, it can be made level when you install your foundation. Also avoid a location where the tank may end up sitting on two different types of materials like cement (i.e., a sidewalk or driveway) and dirt—the dirt could end up washing away, causing your tank to lean dangerously.

How to Choose a Storage Tank Foundation

Choose your foundation based on the type of rain storage tank you've already chosen:

Rain barrel: Choose a cinder block or patio paver foundation. Before installing the rain barrel, place two cinder blocks or a stack of patio pavers on the footprint of the barrel so that the rain barrel will be 8" off the ground. Then place the barrel on top. Cinder blocks and patio pavers are hard materials and will sufficiently distribute the weight of a rain barrel without caving in. Expect to pay around $3 per cinder block or $2 per patio paver brick. Total costs for a rain barrel foundation are around $5 – $10 per barrel foundation.

Cinder blocks are positioned beneath a rain barrel to support the weight of the water and to elevate the barrel.

IBC tote: Place cinder blocks beneath the IBC tote along the perimeter of the tote and then place the totes on top. Patio pavers are *not* recommended for IBC totes because they don't usually have as high of a surface area as cinder blocks do and so are more prone to slipping. That said, pavers can be a good option to place beneath the cinder blocks to add an extra layer of support. Total costs for an IBC tote foundation are around $20 – $80 per tote foundation.

Cinder blocks underneath an IBC tote keep the tote from sinking into the garden.

Cistern (any tank or cistern holding more than ~350 gallons): Cinder blocks and patio pavers underneath tanks with a greater footprint can slip, so it's best to use a compacted sand or gravel foundation or a concrete foundation. That kind of foundation may also be more cost-effective. Sand, gravel, and concrete are the most durable, have the longest lifespans, and cost the least in terms of foundation types. These bases should extend 6" – 24" out from the footprint of your rain tank. For sand or gravel, consider building a frame with two-by-fours of treated wood or rocks, filling it with 4" – 6" of sand or gravel, and then compacting it. You can compact it with a hand tamper tool, or you can rent a vibratory plate compactor from a hardware store.

Sand: While sand is cheap, it can be easily washed away during a storm. If you decide to go with sand, expect to inspect and possibly replace the sand every few months. Sand is recommended *only if* cost is a factor. An alternative to sand is crusher dust (also known as stone dust), which is recycled crushed rock or concrete used for paving. Sand and dust may both be available and are of similar cost and quality. Bagged sand and dust can be found at hardware stores and cost about $5 per cubic foot.

Gravel or crushed rock: Gravel and crushed rock are very common base materials and are best for those who have tighter budgets but also want their base to not wash away as easily. The best types of gravel to use for foundations are pea gravel and crushed rock. Both gravel and crushed rock are commonly used, are inexpensive, and will not wash away as quickly as sand does. The particles are also small enough to prevent them from shifting over time. Gravel and crushed rock are more expensive than sand but will likely last much longer. If cost is an issue, you can mix gravel and sand. Bagged pea gravel and crushed rock can be found at hardware stores for $5 – $6 per cubic foot.

A combination of gravel and stone is used for these tanks' foundation.

Concrete: A concrete base is the most expensive and the longest-lasting option of these three foundation materials. It's also the most difficult to install on a DIY basis. If you've constructed concrete bases before and have the equipment to do so, feel free! Just make sure that the concrete is reinforced with rebar or mesh to keep it strong over time. If you've never installed a concrete base before, consider sticking with compacted gravel (a great option) or hiring a concrete professional to lay the concrete base. A concrete base will cost about $4 – $6 per cubic foot if you do it yourself or about $6 – $8 per cubic foot if a professional lays it.

This metal slimline tank uses a concrete pad for its foundation.

TABLE 12. COST OF EACH TYPE OF FOUNDATION

Type of rain tank	Cost of a cinder block/patio paver foundation	Cost of a sand foundation	Cost of a gravel/crushed rock foundation	Cost of a concrete foundation
Rain barrel	$5 – $10	$80 – $90	$90 – $100	$100 – $130
IBC tote	$20 – $80	$250 – $300	$300 – $350	$200 – $400
500-gallon cistern, 13 ft^2 footprint	Not recommended	$260 – $320	$320 – $375	$200 – $420
2,500-gallon cistern, 35 ft^2 footprint	Not recommended	$700 – $750	$840 – $900	$560 – $1,100

Based on budget and ease of installation, determine which foundation is best for the rain storage tank you've chosen.

4. PIPES OR HOSES (INLET, OVERFLOW, OUTLET)

It can be best to wait to measure and purchase the inlet, overflow, and outlet pipes until after the tank has arrived (or use the tank description to determine these sizes). Also, I recommend placing the tank where it will be located next to your chosen downspout and then measuring the distance needed with a tape measure (see Chapter 6: Install Your System). Consider that distance to be your necessary inlet pipe length. Inlet piping is usually the same diameter as your downspout. Measure this size and plan to purchase a flexible plastic or metal pipe that will attach to your downspout. See Chapter 6 for ideas and illustrations. Overflow piping usually looks like an elbow fitting that fits onto the overflow outlet plus a flexible drain pipe attached to the elbow pointing towards the direction where you want the water to go. Flexible plastic elbows need to have the same diameter that the overflow outlet on your rain storage tank has.

This rainwater tank has an overflow outlet at the top left, running overflows into a pipe system that takes the rainwater to a useful place. This tank is highly elevated for non-powered water pressure.

Measure the diameter of this overflow outlet, then visit the plumbing section of your hardware store to find flexible downspout tubing. Overflow tubing should be long enough to reach from the top of the tank to where you want the overflow water to be routed. Again, it can be best to wait until *after* the tank has arrived to measure and purchase the overflow pipe.

Outlet spigots and hoses will need to be the same diameter as the outlet opening on the storage tank. For rain barrels, simply attaching a spigot to the bottom outlet valve will allow you to fill a watering can or bucket. You can also attach a hose to the outlet valve. Most prefabricated rain barrels have an outlet opening of 5/8" or ¾", which fits most hoses in the US. If you're fabricating your own barrel(s), to create an outlet opening, drill a ¾" or 5/8" hole (depending on the size of your hose) about 4" above the bottom of the barrel to keep sediment from entering the outlet.

One of the advantages of IBC totes is that they already come with an outlet installed at the bottom. Most IBC totes have an outlet size of 2", but measure the diameter of the outlet anyway in case you decide to attach other pipes or fittings to the IBC outlet.

For rain tanks or cisterns, common outlet sizes include ¾", 1", and 2". Many tanks come with a bulkhead fitting already installed in the outlet hole. This is threaded and allows a spigot or pipe manifold to be screwed on to connect to other pipes.

This tank came with a bulkhead fitting already attached. The owner then threaded a pipe manifold into the bulkhead fitting to control the flow of water out of the tank.

Most prefabricated rain barrels, tanks, and cisterns (as well as IBC totes) come with an outlet hole already installed, thus not adding to the cost of installing the entire system. Most brass spigots cost between \$20 – \$25. Alternatives to spigots are ball valves or gate valves (which can be brass) that cost around \$20 or PVC versions that cost around \$5. Once your tank has arrived, measure the outlet opening diameter and purchase a spigot and/or hose that matches that diameter size (again, check the plumbing section of your local hardware store).

A rain barrel with a plastic outlet already installed; the owner then threaded a valve and hose thread adapter so that a hose could be attached to the rain barrel's outlet.

The total costs of pipes and hoses are minimal compared to other components of a rainwater harvesting system, but do plan to spend about \$15 – \$100 on these.

5. FIRST-FLUSH SYSTEM (OPTIONAL)

First-flush systems (also called first-flush diverters) are popular. You may be wondering if they're right for your system. They might be, but then again, they can eventually create more problems than they solve. Because of that, it's best to only use them for specific applications.

During a rain event, first-flush systems divert the first five or so gallons off the roof into a different pipe so that the dirtiest water (which may contain bird feces or heavy metals from the roof) doesn't

end up in the rain tank. For those living in highly urban areas or near industrial sites, pollutants can enter the air and land on the roof and gutters, making first-flush devices useful. They're also useful if a rainwater harvester wants to water edible plants (e.g., a vegetable garden) or their animals or if the harvester feels uncomfortable with the quality of their rain and doesn't want to put in the full energy and expense required to treat their water for full potability.

A first-flush system with a tank; water first enters the first-flush pipe (on the left), which fills up first, then enters into the pipe that flows into the tank.

In most suburbs and rural areas, rainwater quality is high. (Water harvested off a roof in the Arizona desert was found to be more pure than distilled water used in surgery![3]) For most outdoor applications—including watering edible plants—rainwater is suitable. For additional peace of mind, harvesters can water the soil directly instead of watering the actual plant. (The soil already contains bacteria.) Watering animals like dogs, chickens, and goats with rainwater is a fantastic use of rainwater that could benefit from a first-flush diverter, especially in urban or industrial areas. But again, if you live in a rural or even in many suburban areas, a first-flush diverter may not be necessary.

First-flush diverters require frequent manual maintenance, which usually involves cleaning out the fine debris that clogs the drainage as often as every time it rains. If the diverters aren't manually maintained, bacteria trapped in them can colonize the water. Most filtration strategies already mentioned (including gutters, downspouts, and tank screens and filters) will sufficiently remove the largest debris, whereas the finer debris will accumulate at the bottom of the tank in the sludge layer. The sludge layer has been shown to collect heavy metals as well,[4] and that's one of the main reasons why people decide to use a first-flush diverter in the first place. *But* it's worth noting that if the sludge layer isn't disturbed, it will help keep heavy metals and finer debris out of the water in the tank without the risk and manual maintenance that a first-flush diverter creates.

Let's say that you *do* live in an urban or industrial area and a first-flush system could be useful (or a first-flush system could provide you with precious peace of mind regardless of your situation). The easiest way to source the materials for a first-flush diverter is through a kit bought online—search for "first flush diverter kit" to see several different types and then shop around. Find one with the same diameter as your gutter downspout (they usually come in 3"-diameter versions or 4"-diameter versions). Most kits will come with a tee junction (where one end fits onto the downspout, one end fits onto the pipe that goes into the rain tank, and the last end goes into the pipe where the first flush of water enters), a ball (which floats on top of the water as it fills the diverter pipe, sealing it off when the pipe is full), an endcap, and a socket that releases the rain in the pipe slowly enough that the first-flush diverter will work but not so slowly that you need to manually empty out the first-flush pipe every time it rains. This socket is what can clog—it needs to be frequently cleaned. Kits cost $40 – $60 each, and a first-flush diverter can take up to 15 minutes of maintenance per rain event.

First-flush systems have their merits, but to keep rainwater harvesting system installation and maintenance simple (and less costly), I recommend that first-flush systems be used sparingly. Dust, organic matter, and other debris may end up in your water, but then again, some families prefer to be exposed to these compounds rather than avoid them.

Is a first-flush system worth it for you? Ultimately you will make that choice; your decision will depend on your personal comfort and risk level. For general guidelines, see the decision tree in Figure 1.

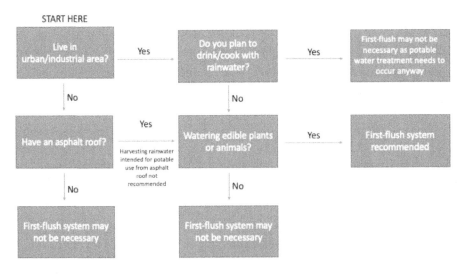

Figure 1. Decision Tree for Determining if a First-Flush System is Recommended for Your Needs

If you decide to purchase a first-flush system kit, know that it won't come with the actual pipe that the first-flush of water will sit in. That's because the length of this pipe is chosen by you. The larger your roof is, the higher total volume of contaminants there will be, and so the longer your diverter pipe needs to be. Some general guidelines are:

- Aim for about 5 gallons of first-flush to go into the diverter pipe for a typical 1,000 – 1,200 ft² roof. After about 5 gallons, you'll sacrifice collecting more precious rainwater to flushing out fewer and fewer contaminants. If you have a particularly large roof, add 5 gallons per 1,000 ft² of roof area.[5]

- A 4" pipe that is about 8 feet long will hold about 5 gallons.

- A 3" pipe that is about 13 feet long will hold about 5 gallons.

Your manufacturer may have different guidelines for use. Follow your kit manufacturer's guidelines.

A diverter pipe holding 5 gallons of water will weigh about 40 pounds. Consider supporting the diverter pipe on the ground along with the cistern foundation and use an elbow pipe at the bottom of the diverter pipe to direct the first flush of water away from any foundations. (Bonus points if the first flush of water is diverted somewhere useful!) All of these fittings and pipes can be purchased at your hardware store. Simply purchase one that has the same diameter as your downspout.

6. PUMPS (OPTIONAL)

Pumps are not necessary, especially for small or simple systems like a rain barrel in the backyard. Pumps can be avoided altogether if the tank sits on a tall hill or if you build a water tower. If you don't have that luxury, however, pumps can be a lifesaver! They can make using rainwater much easier and will incentivize you to use your rainwater. The world of pumps is wide and deep, but I'll recommend a few pumps that have been used by other rainwater harvesters for their cost, ease of installation, and ease of use.

There are two main types of pumps to consider as a rainwater harvester: submersible pumps and external pumps. Submersible pumps sit inside the rain tank and are submerged in the water they pump; external pumps sit outside the rain tank and are connected to the rainwater via the tank's outlet.

A submersible pump sitting in the cistern where it pumps water from.

An external pump (with a cover over it) sitting outside the cistern from which it pumps water.

External pumps need to have initial water poured into the pump (this is called "priming the pump," which you'll need to do unless you find that the pump "self-primes" in the product description), whereas submersible pumps don't need this. Submersible pumps are quieter, but unlike external pumps, submersibles need to be removed from an aboveground tank before winter if the aboveground tank might be threatened by frost. Submersible pumps can also be harder to service and replace if they are placed in large tanks. In general, submersible pumps can be longer-lasting and more expensive than external pumps are, but that's a generalization. You'll find plenty of exceptions to that.

Pumps come in manual versions (i.e., you turn it off and on yourself when you need and don't need the water) and automatic versions (i.e., it turns off and on to maintain a certain pressure or flow rate or turns off and on at certain times of the day). Generally, manual pumps are less expensive than automatic pumps. You may also see pumps at the hardware store called "transfer pumps." Most transfer pumps are manual. They just need to be turned off when not in use.

Pump selection can be complicated, but it's possible to simplify it. Which pump you choose will depend on your application, budget, power needs, flow rate and head pressure need, and personal preference. I encourage doing your own research to figure out which pump would be best for you. Otherwise, based on prior input from other rainwater harvesters, my calculations, and professional recommendations, Table 13 provides guidance on shopping for a pump depending on your application. The manufacturers listed aren't necessarily recommended but are meant to provide guidance. Pumps were chosen based on how available they are online and at common hardware stores like Harbor Freight or Home Depot. If you're purchasing a large, expensive pump online, it's best to purchase from a reputable dealer rather than a more generic business like Amazon. That's because specialized dealers can better assist you with pump selection, plus great care must be taken during shipping and handling to avoid the pump being damaged.

TABLE 13. OPTIONS FOR PUMPS DEPENDING ON APPLICATION

Application	Low-Cost Option ($100 – $500)	High-Cost Option ($200 – $800)
Outdoor (drip irrigation)	Drummond 1/10 HP Transfer Pump at Harbor Freight	Shurflo 2088-594-144 Industrial Automatic Pump Demand Diaphragm Pump on Amazon
Outdoor (sprinklers, hose-end sprayer nozzle)	Drummond 1 HP Portable Sprinkler Pump at Harbor Freight	Leader Divertron ¾ HP Submersible Pump on RainHarvest.com
Indoor (toilet, faucet, washing machine, dishwasher); *submersible pump preferred*	Leader Divertron ¾ HP Submersible Pump on RainHarvest.com	Strom 1 HP Automatic Bottom Suction Cistern Pump on Plastic-Mart.com
Indoor (toilet, faucet, washing machine); *external pump preferred, pressure tank included*	Drummond 1 HP Stainless-Steel Shallow Well Pump and Tank with Pressure Control Switch at Harbor Freight	Red Lion RL-SWJ50/RL6H ½ HP Shallow Well Jet Package, Cast-Iron Pump with Pressure Tank, 5.8 Gallon at Northern Tool (useful for off-grid homes)

When you shop for a pump, look for shallow well pumps, transfer pumps, jet pumps, or booster pumps (see Pump Glossary for explanations of these terms). These pumps are usually used to pump wells or move clear water from one place to another, but they also work well for moving rainwater. When you buy your pump, you'll also need to purchase a check valve or a foot valve. (These are valves that keep water from backflowing into the pump.) Check with your manufacturer to find out which type and size of valve to get. Exceptions include pumps that come with a check valve already installed.

If your pump doesn't come with a pressure tank (the recommended external pumps in Table 13 already come with a pressure tank) and you plan to use your rainwater for many indoor applications, consider purchasing a pressure tank along with a pump. You'll hook it up to your pump at the pump's outlet, and a pipe from the pressure tank's outlet will feed your faucets or appliances. A 20-gallon pressure tank can be sufficient for indoor use, and you can purchase one for about $180 – $300 at hardware stores like Home Depot or Lowe's. A pressure tank will keep the water under pressure so that the pump doesn't have to work overtime. It will increase the lifespan of your pump!

Pumps generally have a fixed power need: in the US, pumps usually come in ½ HP (horsepower), ¾ HP, 1 HP, or any variation thereof. As you shop around, pay attention to the horsepower level the pump needs and the specified voltage and amperage for the pump (these specifications are in the pump description on the box or online). The horsepower of the pump will determine two things: its flow rate (how *fast* the pump will push water through it) and its head pressure (how *hard* the pump will push water through it up a vertical distance). A pump with 1/10 HP – ½ HP will work well for most outdoor and garden applications, and a pump with ¾ – 1 HP will work well for most indoor applications. (Note that your particular needs may vary.) Indoor appliances in the US can need between 40 to 100 psi of pressure. Your pump and pressure tank need to match this pressure! Otherwise, you'll need to choose appliances with fixtures that can accommodate lower pressures.

7. INDOOR-USE FINER FILTRATION

If you plan to use your rainwater for your garden or outdoor use in general, you can skip this section. If you plan to use your rainwater for indoor purposes, read on. Just as many varieties of pumps exist, you'll also find many filtration and treatment options for your rainwater harvesting system. The most essential and most commonly used ones will be covered here, but I encourage you to do your own research as well. If you don't plan on drinking your rainwater, filtration needs to be about 100 microns or less. Purchase filtration systems from hardware stores like Home Depot or Lowe's (look for water or well water filtration systems that filter between 10 – 30 microns). For potable water, add on even finer filtration, between 5 – 20 microns.

Note that the higher the filtration is, the more slowly the water will flow and the more expensive the filtration will be. Filtration cartridges will likely need to be replaced annually.[6] A finer level of filtration will remove sediment and debris that your gutter and tank screens didn't filter out. See Chapter 6: Install Your System to see examples of cartridges and filters that can be used indoors.

Examples of water filters you can purchase at a hardware store; for drinking water, consider a filter of 5–20 microns.

8. INDOOR-USE PIPES

You'll need to purchase pipes that span the length between your indoor pump and filtration system to your application (e.g., toilet, faucet, appliance). Common pipe materials are PVC, CPVC, PEX, or copper. Given that the plumbing aspects of installing pipes can get extremely granular and specific, I'll only cover the need for them here. Specifications for your home lie in the plumber's realm and probably deserve the attention of an entire (other) book! A great resource for learning how to DIY your own plumbing is *Ultimate Guide: Plumbing* by the Editors of Creative Homeowner. It makes a great companion to this book.

In the meantime, you'll want to choose *one* type of material for the plumbing. PVC cannot carry hot water and is known to negatively impact the environment, but it's very easy to find and work with.

CPVC can handle temperatures up to 200°F and has similar mechanical properties to PVC. PEX is a cross-linked polyethylene and can handle hot water better than PVC or CPVC can. PEX is also very flexible, but it's more expensive and special tools are needed to work with it. PEX piping is also compatible with copper pipes if you have those in your home. Copper is expensive but long-lasting and can handle most water temperatures. All of these pipes can be purchased at a hardware store like Home Depot or Lowe's or at your local plumbing supply store. For your specific project needs, you may need to consult a plumber. Tell them you plan to hook up your appliances to your rainwater harvesting tank and they will be very happy to help!

9. POTABLE WATER TREATMENT

If you do plan on drinking your rainwater, add a 5 – 10-micron filter right after your 10 – 30 micron filter. (These filters can be purchased from hardware stores like Home Depot or Lowe's or from specialty rainwater harvesting stores like Rainharvest.com.)

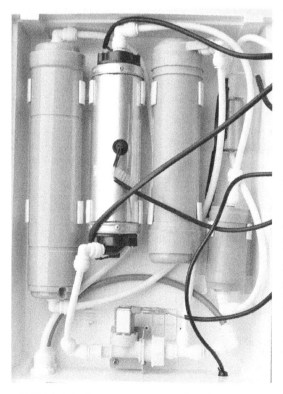

To treat your rainwater for disease-causing microbes, consider several different modalities, such as countertop potable water filters like the ones made by Berkey, disinfecting using ozone, disinfecting using UV light, and disinfecting using chlorine. (You have many other options, but these are the four I'll be covering.)

Each type of filtration has its pros and cons and its lovers and haters. The lower the startup costs for a filtration system are, the higher the per-gallon costs of filtration generally are over the long run due to having to replace the filtration components much more often. I encourage doing your own research—only you can determine which options and levels of risk best suit your situation. Use Table 14 to determine which method could be right for you.

A UV lightbulb treatment setup with in-line water filtration components.

TABLE 14. RAINWATER DISINFECTION METHODS AND THEIR ASPECTS

Rainwater Disinfection Method	Startup Costs	Method of Disinfection	Pros	Cons	Where to Purchase
Countertop filtration systems like Berkey	$200 – $400	Uses filters that filter out microorganisms	Quick and simple startup solution; common and easy to find; filters need to replaced every 2 – 5 years	Expensive long-term solution; filters only up to 6 gallons at a time	Online at Amazon or BerkeyFilters.com; from hardware stores like Home Depot
Ozone disinfection	$250 – $400	Injects ozone into water, which oxidizes and kills microorganisms	Powerful and reliable; able to treat water even if the water is cloudy	Hard to find in US at economical prices; ozone leaks in the home can be dangerous; requires ozone-resistant materials (e.g., stainless steel)[7]	Rainharvest.com

Rainwater Disinfection Method	Startup Costs	Method of Disinfection	Pros	Cons	Where to Purchase
UV disinfection	$600 – $1,000	Shining UV light through water to render microorganisms unable to reproduce	Leaves behind no byproducts; easy to install and maintain; because it's a fairly common method, it's easier to find bulbs	Water must be highly filtered first or else light won't penetrate through to the microorganisms; expensive; bulbs need to be replaced about annually	Rainharvest.com; Pelicanwater.com; some hardware stores like Home Depot; water treatment companies and stores
Chlorine disinfection	$30 – $50	Adding chlorine bleach straight to the water (2.3 fluid ounces per 1,000 gallons of water)[6]	Inexpensive; easy to implement straight away; easy to scale	Suspected negative side effects due to byproducts of trihalomethanes (such as chloroform)	Purchase chlorine bleach at any grocery store or hardware store

Phew! You made it! Now it's time to take action.

ACTION GUIDE FOR STEP 5: SCALE AND SOURCE YOUR SYSTEM

1. Now that you've determined your end-use goals, the amount of rainwater you'll be able to (and want to) collect, and the amount of water you'll need for your applications, decide on how big your tank will be. This step is what's called "scaling your system." Every system described in this book is just a bigger version of a rain barrel. Those only wanting to water a garden may use a 55-gallon rain barrel, whereas those wanting to set up their first water source for their off-grid cabin may choose a 5,000-gallon rain barrel known as a tank or cistern. Choose your tank size based on your budget, your water needs, the amount of water you can collect, and your space availability. You can always add more tanks later, but it's usually harder to downsize. That said, if you can afford it, don't be afraid to go for a larger tank! That can help you store more water during droughts or dry spells. But also don't be afraid to try out a smaller size to build your confidence in installing your first system. You can always try something else or something new later on. Either way, note that how you scale your system will determine the size of your system.

2. Decide which type of tank you would like to install and where you'll buy it or order it from (using Tables 9, 10, and 11).

3. Having decided which tank you need and want, decide on which type of foundation you need and want and where you'll source the materials (using Table 12).

4. Decide if you would like to use/need a first-flush system (using Figure 1).

5. Decide if you would like to use/need a pump. If you would, choose which type of pump you would like to use and where you'll buy it or order it from (using Table 13).

If you plan to use your water indoors, determine which type of indoor filtration and treatment system you'll use and where you'll buy it or order it from. Consult external sources like plumbers or plumbing guidebooks if you need help with making indoor plumbing decisions. If you're not used to doing your own plumbing, head to your local plumbing supply store or hardware store to speak to a plumbing professional. Explain to them that you want to hook up a rainwater harvesting system to your faucets and appliances. They'll be delighted to steer you towards what types of pipes and fixtures to purchase. They can also explain the specifics of how you'll connect your tank to your appliances.

6

If you plan to use your rainwater for potable use, determine which modality of potable water treatment you and your family wish to use and where you plan to purchase it from (using Table 14).

7

When your tank arrives, measure the sizes of the inlet, overflow outlet, and outlet openings to determine what sizes and lengths of inlet pipes, overflow pipes, and outlet spigots and hoses you will need. Determine where you'll buy them. (These components are usually available at hardware stores like Home Depot or Lowe's.)

8

INSTALL YOUR SYSTEM

CHAPTER 6

Once you've acquired your materials, you can start installing your very own rainwater harvesting system! The installation guides in this book will walk you through the steps for installing a rain barrel, IBC tote, and cistern. Again, these are basically all varying sizes of a rain barrel.

1. Pick a downspout
2. Install the foundation
3. Place the storage tank on the foundation
4. Connect the downspout to the storage tank
5. Route the overflow away from the adjacent building
6. Configure the outlet and enable the rainwater to be used

Each of these systems comes with variations that could be useful for you, such as being able to connect the downspout to a pump, drip irrigation system, or filtration system. Here, you'll find installation guides for these variations, including:

- How to connect two or more tanks together
- How to install a first-flush system
- How to install an automatic rainwater chicken-watering system
- How to connect a storage tank to a toilet (or any indoor appliance)

As is the case with any guide, read the entire guide before buying any materials. This could save you a few trips to the hardware store! All of the setups explained in these guides are meant to illustrate and inform, and all can be configured to match your personal preferences, creativity, expertise, and budget. Keep the foundations of safety and use in mind and ensure that all storage tanks have not only an inlet and outlet, but also an overflow outlet and vent.

If you're new to the world of plumbing, see the Plumbing Glossary at the back of this book to become more familiar with specific terms. Also, you'll want to ask for individual guidance from experts at plumbing stores and hardware stores. Also consider purchasing DIY home plumbing books like *Ultimate Guide: Plumbing* by the Editors of Creative Homeowner.

HOW TO INSTALL A RAIN BARREL

This installation guide is the simplest, most beginner-friendly system in the book. You'll use a prefabricated rain barrel (one you can purchase from a store or online) that already has a built-in screen, overflow outlet, and spigot with threads for a garden hose. You'll only be using the screen on the barrel to filter the rainwater. This system is best for pure beginners, seasonal gardeners, and outdoor non-potable rainwater users who just want to collect rain sparingly without too much investment. These instructions also include how to cut a downspout. If you don't want to fully cut off the downspout (especially if you're installing a seasonal system in a freezing climate), see the Variations section of the guide for alternatives.

Skill level: Beginner

Approximate time: 60 – 120 minutes

Equipment needed:

- Bubble level
- Tin snips or hacksaw
- Tape measure
- Electric drill
- Permanent marker

Materials needed:

For the foundation:

- (2) cinder blocks

For the storage tank:

- (1) prefabricated rain barrel with inlet screen, overflow outlet, and spigot with threads for a garden hose thread (example: RTS Home Accents 50-Gallon Rainwater Collection Barrel with Brass Spigot, available on Amazon)

For the downspout and inflow:

- (1) flexible downspout elbow (extending between 8" – 18")
- (4) downspout screws or sheet metal screws (between 3/8" – ½")

For the overflow:

- (1) pipe or flexible pipe that's the same diameter as the overflow outlet on the rain barrel

For the outlet:

- (1) hose

Instructions:

1. Pick a downspout

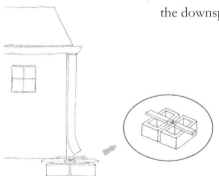

    1) Locate all downspouts on the house.

    2) Pick a downspout. Considerations when choosing a downspout:

        a) Note the ground underneath each downspout. The more level the ground is, the better. The area underneath the downspout should also have enough space for the rain barrel to sit on the ground.

    b) For gardeners, the closer the barrel is to the garden bed/wherever the water will be used, the easier it will be to bring water to the plants.

    c) Note which downspout could be the easiest to modify. All of your downspouts may look the same, but perhaps one is easier to access. Also consider the materials—aluminum gutters are easier to cut than, say, steel.

2. Install the foundation

    1) Clear out any rocks or pebbles on the ground directly underneath the downspout.

        a) Once the area has been cleared, use a level to measure how level the ground is beneath the downspout. Using a shovel or rake, move the dirt around to make it level.

    2) Place the cinder blocks underneath the downspout with its open side facing up. The footprint needs to be as wide as the rain barrel is.

3. Place the storage tank on the foundation

1) Place the rain barrel on the cinder blocks right underneath the downspout. Adjust the cinder blocks if needed to make them more level.

2) Place a level on top of the rain barrel to check if it's level. If it's not, remove the rain barrel and adjust the level of the dirt. **Do not proceed until the system is level.** When the rain barrel fills, it could tip to one side and fall over if it's not level.

4. Connect the downspout to the storage tank

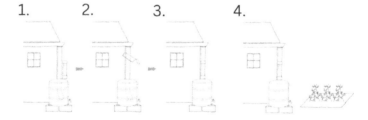

1) With the rain barrel on top of the cinder blocks and directly underneath the downspout, measure between 8" – 15" above the rain barrel on the downspout. Make a mark on the downspout. This is where the downspout will be cut and you'll attach a flexible elbow to the downspout and point it into the inlet of the rain barrel.

2) Using tin snips or a hacksaw, cut the downspout to the correct length.

3) Attach the downspout side of the flexible elbow to the downspout using screws.

4) Point the other end of elbow into the inlet on the rain barrel. If you need to point the elbow at an angle, consider using pipe straps to keep the elbow in place alongside the adjacent building.

5. Route the overflow away from the adjacent building

1) Measure the diameter of the overflow outlet on the rain barrel (your rain barrel may come with specifications of the overflow outlet size).

2) Measure the distance away from the building that the overflow outlet should reach. When the rain barrel overflows, the overflow should move water safely away from any foundations and preferably to somewhere useful, like a garden bed.

3) Purchase flexible pipe that meets these specifications. Attach one end to the overflow outlet and point the other end away from the adjacent building.

6. Configure the outlet and enable the use of rainwater

1) Test to see if the spigot is working by pouring water into the rain barrel.

2) After the next rain, place a watering can underneath the spigot and bring the rainwater to your garden bed. Alternatively, attach a hose to the spigot.

You're done! Congratulations!

Variations on this setup:

- For those who aren't keen on sawing off an entire downspout or who would rather flip a switch to winterize their system, use a downspout diverter. Diverters work with the downspout to divert rainwater into a barrel so that the original downspout stays mostly intact. The downside of diverters is that your catchment efficiency will be lower. Some examples of diverters are the Oatey Mystic diverter and the EarthMinded DIY Rain Barrel Diverter and Parts Kit (both are available on Amazon). Also see the RainHarvest Clean Rain Ultra, an all-in-one filter and diverter.

  Diverters allow for quick takedown before the winter months; diverters also route precipitation back down the main downspout when freezes start to set in. To use a diverter, replace Step 4 with instructions from your diverter or diverter kit. A combination of sawing off the downspout and using a diverter is illustrated in the upcoming How to Install an IBC Tote section.

- Instead of buying a prefabricated barrel, create your own by using a recycled car wash soap barrel, an olive barrel, a whiskey barrel, or any variation thereof (the list goes on and on). There are tons of instructions online on how to convert a regular barrel to a barrel with inlets, outlets, and overflow outlets. In addition to being a diverter kit, the EarthMinded DIY Rain Barrel Diverter and Parts Kit can be used to convert almost any barrel into a rain barrel.

- Attach a drip irrigation system to your rain barrel by using a drip irrigation kit like the Mister Landscaper Drip Irrigation Vegetable Garden Kit (available at Lowe's). Follow the manufacturer's instructions on how to attach the drip irrigation kit to a spigot. Instead of attaching it to a spigot on the side of an adjacent building, attach it to the spigot at the bottom of your new rain barrel.

- Cinder blocks aren't necessarily the most aesthetically pleasing base. Consider installing a gravel pad with pretty pea gravel. Rain barrels are also often supported with treated wood platforms, so those could be an option if you have experience with building them or if you find that you can purchase one. Otherwise, stick with cinder blocks or gravel.

- If you need to bring water from the rain barrel to a garden that's up a hill, consider using a small manual pump such as a ½ HP (or less) transfer pump. Use a hose to connect the outlet of the rain barrel to the inlet of the pump and then use another hose at the outlet of the pump to bring the water up the hill. You can automate watering with a timer that can be purchased at a hardware store.

This setup is the simplest and doesn't include any gutter downspout diversions. If you live in a freezing climate, before winter hits, you'll need to drain the barrel and no longer point the flexible elbow into the barrel. Ideally, point it down the length of the building again and into a downspout. See the Variations section for options for making winterization easier.

HOW TO INSTALL AN IBC TOTE

This setup is inspired by Jeff Trapani, @nature_hacker. This setup uses a downspout diverter, which can be used in any of the guides in this chapter. The one used in this setup combines filtration and diversion (an example is the Rain Harvesting Clean Rain Ultra, available online at RainHarvest.com) and was chosen because it's simple, easy to adapt to most downspouts, and simple to winterize. It uses a manual switch that diverts water either to the storage tank or down the main downspout. This kind of system is particularly useful in freezing climates—you can just flip the switch before the first deep freeze hits (and empty out any contents of the tote and pipes). This setup is configured to allow you to attach a ¾" garden hose with FHT threads to the IBC tote's outlet. If you need to attach any other size hose or pipe, go with the sizes you need. IBC totes are useful because they're cheap and easy to find, and they hold a fair amount of water. However, they won't come animal/insect/child-proof as-is. Also, if you have a translucent tank, you'll want to paint the tank or cover it to render it opaque. Remove the metal cage, then paint or cover the tank, then replace the cage after you're done. (The cage provides support for the tank.) Paint the tank with several coats of plastic spray paint such as Rust-O-Leum Plastic Spray Paint and let it cure fully. Alternatively, cover the tank with black plastic sheeting.

For all PVC pipes, spray-paint them as well with a similar plastic spray paint and let the paint cure fully. This will prolong the lifespan of the pipes. Plan to replace them when they get brittle, as they can crack and cause leaks.

Skill level: Intermediate

Approximate time: 2 – 4 hours

Equipment needed:

- Bubble level
- Shovel/rake
- Tape measure
- Tin snips or hacksaw
- Electric drill

- Pliers or downspout crimper
- Sandpaper
- Teflon tape
- Silicone caulk

Materials needed:

For the foundation:

- (8 or 16) cinder blocks
- (5) bags crushed rock (optional)

For the storage tank:

- (1) IBC tote, already painted opaque or covered in opaque sheeting

For the downspout and inflow:

- (1) diverter (e.g., Clean Rain Ultra from RainHarvest.com)
- (1) IBC cap with 2" threaded hole (e.g., Banjo 6 ½ IBC Cap with 2" Threaded Plug Hole, available on Amazon)
- Paint to paint all PVC components
- (1) 2" PVC male adapter
- (1) 2" PVC sanitary tee
- (1) 2" PVC 90° elbow
- (1) 2" PVC straight pipe (length determined by your measurements but at least 23")
- (1) 3" to 2" PVC reducer bushing

For the overflow:

- (1) 2" PVC straight pipe (length determined by your measurements but at least 23")
- (1) 2" PVC 90° elbow
- (1) 2"flexible corrugated tubing (optional)

For the outlet:

- (1) 2" PVC coupler
- (1) 2" to ¾" PVC reducer bushing
- (1) ¾" MIP to ¾" MHT adapter nipple
- (1) ¾" hose

Instructions:

1. Pick a downspout

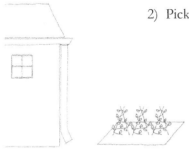

    1) Locate all downspouts on the house.

    2) Pick a downspout. Considerations when choosing a downspout:

        a) Note the ground underneath each downspout. The more level the ground is, the better. The area underneath the downspout should also have enough space for the IBC tote (at least 45" x 45", or 14.06 ft^2).

        b) The closer the downspout is to where the water will be used, the easier it will be to bring water to where you need it.

    c) Note which downspout could be easiest to modify. All of your downspouts may look the same, but perhaps one is easier to access. Also consider the materials—aluminum gutters are easier to cut than, say, steel.

2. Install the foundation

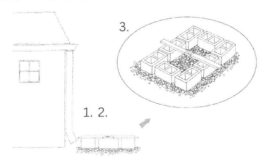

    1) Clear out any rocks or pebbles on the ground directly underneath the downspout.

    2) Once the area has been cleared, use a level to measure how level the ground is beneath the downspout. Using a shovel or rake, move the dirt around to make it level.

    3) Place cinder blocks at least 23" to the right or left of the downspout. This is a little more than half the width of the IBC tote; you'll be connecting the downspout to the tote inlet in the middle of the tote. Measure a square shape around the perimeter of the footprint where the tote will be positioned. It can be useful to use a tape measure to mark out the footprint of the tote first, then place the cinder blocks.

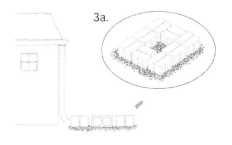

a) If you find the area to be rather soft or muddy, it can be helpful to place the cinder blocks with their closed sides up (the holes or openings would be facing outwards) and place two cinder blocks next to each other where there would be only one if its closed sides weren't facing up. Doing this will decrease the surface area over which the full tote will exert its force, lessening any potential sinking effects. Alternatively, instead of doubling the amount of cinder blocks, you could cover the muddy ground with crushed rock (this comes in bags at hardware stores) or a base of coarse stone pavers to create a harder surface over which the force of the tote can be distributed.

4) Use a level to double-check that the base is level.

3. Place the storage tank on the foundation

1) Place the IBC tote on the cinder blocks. As the tote inlet is in the middle of the tote, the tote inlet should be at least 23" to the right or left of the downspout.

2) Place a level on top of the tote to check if it's level. If anything, try to point the tote away from the adjacent building so that in the event that the tote tips, it tips away from the building and away from the foundations.

4. Connect the downspout to the storage tank

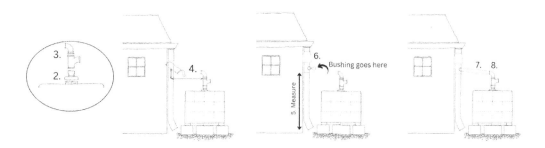

1) This setup uses an all-in-one downspout diverter and filter. You may choose to use a different filtration or diversion method—the principles are the same.

2) To make the IBC tote animal/insect/child-proof, you'll need a way for water to get inside without animals or critters also getting inside *and* you'll need to let air escape as the tote fills. To achieve this scenario, this setup will use a special IBC tote lid that has a 2" opening at the top. Remove the lid that the IBC tote came with and screw in the new lid.

3) Screw the 2" PVC male adapter into the new IBC tote lid. Then connect the 2" sanitary tee into the male adapter with the horizontal end facing the direction where you want the overflow to go. On top on the sanitary tee, place the 2" 90° elbow with the elbow facing the downspout. You'll need to do all of these things first, as this setup will determine how high up you'll need to cut the downspout so that you can insert the diverter.

4) Eventually, you'll attach the horizontal 2" PVC pipe to the end of the elbow. This will carry water from the diverter to the IBC tote. However, you'll only do this *after* you've measured how long the pipe needs to be. Before you install the diverter, estimate the elevation change the horizontal pipe will need to take on to enable the water to travel down the horizontal pipe. As long as the slope of the horizontal pipe is about 1/8" per foot of pipe[1], that's sufficient (a greater slope is fine).

5) Measure up the downspout to where the diverter will be placed based on the vertical distance measurements you made. Use the manufacturer's instructions to also guide you. Measure the vertical distance on the downspout where you'll need to cut off the downspout. Using tin snips or a hacksaw, cut off the section of the downspout where the diverter will sit. Install the diverter in between the remaining sections of the downspout following the manufacturer's guidelines. Crimp any edges of the downspout to the diverter, then swivel the diverter's diversion end towards the IBC tote.

6) The diversion end diameter will likely be 3". Using the 3" to 2" PVC reducer bushing, place the 3" end into the diverter end with the 2" end sticking out. You will be pushing your horizontal pipe into the bushing.

7) Measure the distance between the end of the bushing and the end of the elbow that's attached to the tote. As these parts may need to be inspected and cleaned from time to time, it's best to *not* connect all of these parts with PVC cement. Instead, make sure the pipe is long enough so that it can sit between the bushing and the elbow without cement. Using a hacksaw, saw off a 2" PVC pipe of that length. Use sandpaper to sand off any burrs.

8) Attach one end of the 2" PVC horizontal pipe to the 3" to 2" PVC reducer bushing, then attach the other end to the elbow.

5. Route the overflow away from the adjacent building

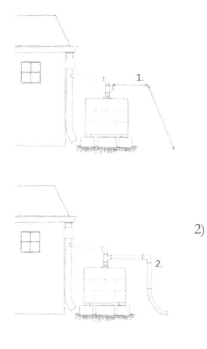

1) Measure a length of horizontal pipe that's long enough to route any overflow away from the tote (perhaps just long enough to reach outside the IBC tote). Cut a 2" PVC pipe to length and sand off any burrs. Connect this horizontal pipe to the other end of the sanitary tee. The sanitary tee is useful as it works well to divert vertical flow in a horizontal direction when the tote overflows. Measure the angle of the overflow pipe and try to point the pipe down so that gravity can assist the overflowing water out into the horizontal overflow pipe.

2) Connect the horizontal overflow pipe to a vertical 2" PVC pipe, using a length that properly diverts water away. Connect the horizontal and vertical pipes with a 2" PVC elbow. At the end of the vertical pipe, you could connect a 2" flexible downspout to continue to divert water to a garden bed or away from the tanks (or leave as-is).

6. Configure the outlet and enable the use of rainwater

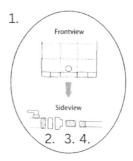

1) Your IBC tote will likely come with an outlet opening at the bottom. The diameter of this hole varies, so double-check its diameter. You'll need to wrap Teflon plumber's tape counterclockwise two to three times around all of the threads between all of the fittings for the outlet configuration *before* you screw the pieces together.

2) The outlet diameter will likely be 2". Screw on a 2" PVC coupler, followed by a threaded 2" to ¾" PVC reducer bushing.

3) Next, screw in a ¾" MIP to ¾" MHT garden hose adapter.

4) This is where a hose can now screw onto the end. Open the valve on the IBC tote to open the system. This lets water flow into the hose when the tote starts to fill.

5) Before leaving the site, be sure that the lid on the IBC tote is not too tight. This will allow air to escape as the tank fills. Test the system by pouring water into the diverter (filling it with a garden hose works well) and checking for leaks. Leaks can be mended with silicone caulk or more Teflon tape, although note that too much caulk or tape can lead to more leaks.

Congratulations! You're done!

Variations on this setup:

- Instead of using a diverter, replace the diverter with a downspout filter (such as a Leaf Eater or similar filter) and place the IBC tote directly underneath the downspout. To winterize a system like this, you can use a Y downspout and divert your water to the Leaf Eater during the summer, then manually switch the Y over to the downspout when winter arrives. (You'll also have to empty the tank.)

- For a more aesthetically pleasing foundation, instead of cinder blocks, install a gravel pad underneath the IBC tote with pretty gravel surrounding it.

- Attach a drip irrigation kit or pump to the end of the garden hose adapter to have more options for watering the garden or bringing water into the home. If you would like to attach a garden hose nozzle to the end of the hose, attach a ½ HP transfer pump between the end of the MHT outlet and the garden hose.

HOW TO INSTALL A CISTERN

This setup is inspired by the Brotherton family at Better Together Life, @bettertogetherlife. It's applicable to any large cistern, from 500 gallons to about 5,000 gallons. As long as your site has sufficient space, play around with tanks that are as large as you need them to be! This setup uses a gravel pad as well as a basket strainer in the inlet of the cistern. If you'd prefer a gravel pad for your rain barrel or IBC tote, follow these instructions for a gravel pad for your rain barrel/IBC tote setup. And if you wish to use an in-line filter (like Leaf Eater) rather than using a basket strainer as instructed in this guide, see the Variations section. Another feature of this setup is a tank gauge—this lets you keep an eye on the water level in an opaque tank. Tank gauges can be used for IBC totes as well (and even rain barrels).

This setup is configured to allow you to attach a ¾" garden hose with FHT threads to the IBC tote's outlet. If you need to attach any other size hose or pipe, go with the sizes you need.

Skill level: Intermediate

Approximate time: 10 – 12 hours

Equipment needed:

- Bubble level
- Shovel/rake
- Tape measure
- String/spray paint (optional)
- Electric drill
- Hand tamper tool or plate compactor

- Sandpaper
- All-purpose PVC cement
- Teflon tape
- Silicone caulk
- Jigsaw or hacksaw

Materials needed:

For the foundation:

- (4) 2"x 4" pressure-treated wooden beams (the length will be determined by your measurements)
- (20 – 40) bags of gravel, crushed rock, or sand

For the storage tank:

- (1) cistern that comes with preinstalled outlet and inlet ports (an example is the Norwesco 2,500-gallon Plastic Potable Water Storage Tank, available on RainHarvest.com or Plastic-Mart.com)

For the downspout and inflow:

- (1) 3" PVC straight pipe (the length will be determined by your measurements)
- (1) 3" PVC check valve (optional; available online and at plumbing supply stores)
- (1) downspout adapter (if your downspouts are rectangular and/or a different size than 3" circular)
- (4) downspout screws or sheet metal screws ($\frac{3}{8}$" – $\frac{1}{2}$")
- (1) 3" elbow (the angle will be determined by your measurements, ideally <90°)
- (1) 3" pipe strap (optional)
- (1) basket strainer or inlet screen (e.g., 8" basket strainer available online)
- (1) pair of pantyhose or mosquito-proof strainer mesh
- (1) 3" hinged vent flap (optional; also known as a Mozzie Stoppa, available online)

For the overflow:
- (1) 1.5" PVC male bulkhead adapter
- (1) 1.5" PVC 90° elbow
- (1) 1.5" PVC straight pipe (the length will be determined by your measurements)
- (1) 1.5" flexible corrugated tubing

For the outlet:
- (1) 2" PVC male bulkhead adapter
- (1) 2" ball valve

For the garden hose:
- (1) 2" to ¾" reducer bushing
- (1) ¾" MIP to MHT adapter nipple
- (1) garden hose

For the pump:
- (1) pump (external or submersible)
- (1) adapter or bushing (the diameter will be determined by the pump inlet)
- (1) check valve or foot valve (the diameter will be determined by the pump inlet; follow the manufacturer's instructions)
- Pipes/hoses to connect the tank outlet to the pump (see your manufacturer's guidelines for proper specifications)

1. Pick a downspout

1) Locate all downspouts on the house.

2) Pick a downspout. Considerations when choosing a downspout:

- Note the ground underneath each downspout. The more level the ground is, the better. For a large cistern, your options narrow, as you will likely need the downspout that's closest to wherever you have available space for a cistern.

- Note which downspout could be easiest to modify. All of your downspouts may look the same, but perhaps one is easier to access. Also consider the materials—aluminum gutters are easier to cut than, say, steel.

- Survey the site near the downspout where you envision placing the tank. This is where the gravel pad and tank will go.

2. Install the foundation

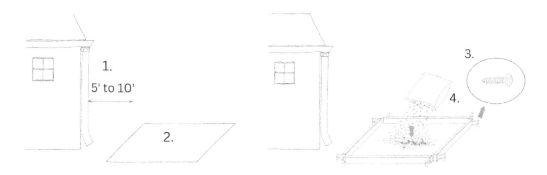

1) To install a gravel pad, first clear the area around the downspout where you plan to place the cistern. Don't plan to place the cistern too far away from the gutter! The greater the horizontal distance is, the less vertical supporting force the pipe will have between the gutter and the cistern. This distance will vary between sites, but you can plan for around 5' – 10' of horizontal distance between the gutter outlet and the cistern.

2) Measure the footprint of the cistern. The footprint of the gravel pad should be 16" – 24" wider than the footprint of the cistern on all sides. Use a tape measure, string, or spray paint to mark the edges of the gravel pad once you've measured these dimensions.

3) The perimeter of the gravel pad can be outlined with pressure-treated 2"x 4" wooden beams to keep the gravel in place. To do this, given the measurements you took of the gravel pad's perimeter, cut 2"x 4" wooden beams to length and screw them together with wood screws. Place the wooden beam perimeter on the edges you marked down.

4) Fill the wooden beam perimeter with 4" – 6" of pea gravel (or depending on your personal preference, sand or crushed rock). Using a hand tamper tool, compact the gravel as much as possible. An alternative to a hand tamper is a vibratory plate compactor. (This tool can make the job quicker.)

3. Place the storage tank on the foundation

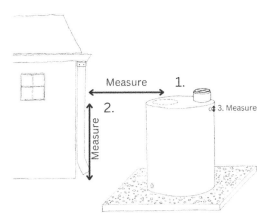

1) Move the cistern into place on top of the gravel pad. This is usually accomplished by pushing the tank onto the top of the pad.

2) Measure the horizontal distance from the bottom of the gutter downspout to the top of the tank. Also measure the vertical distance.

3) Measure the diameter of any overflow outlets. This will typically determine how big your horizontal pipe size will need to be. As a general principle, your overflow port will be a minimum of half of the diameter of the inlet[2]; if your tank has 1.5" overflow outlets, a 3" inlet should do fine.

4. Connect the downspout to the storage tank

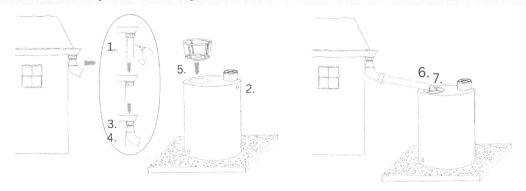

1) You'll first want to completely remove the downspout you chose, as you will next connect the horizontal pipe coming from the gutter to the tank at the base of the gutter. This is usually accomplished by unscrewing any screws that connect the downspout to the gutter as well as unscrewing the pipe straps that keep the downspout flush against the side of the building. You'll be left with the short bit of the downspout where water travels from the gutters down to what was previously your downspout.

2) Your prefabricated cistern may come preinstalled with an outlet, inlet, and (possibly) a manway port. Use the inlet as the overflow outlet/vent and estimate the size of the inlet hole at the top of the cistern. How big your inlet will be is up to you, but plan for it to be about twice the size of your overflow outlet. In this setup, you'll use a 3" pipe (painted PVC Schedule 40, but preferably Schedule 80), as this is the best balance between pipe sizes and their availability at common hardware stores and the pipe being large enough to accommodate a backflush in the case of overflow. As a general principle, your overflow port should be a minimum of half of the diameter of the inlet; on a tank with 1.5" overflow outlets, a 3" inlet should do fine. For extra gutter backflow prevention, install a 3" PVC check valve in line with the horizontal pipe.

3) Connect a rectangular-to-circular 3" downspout adapter to the rectangular downspout with screws. (Note that this setup assumes that you don't necessarily have the ability to screw pipes into the side of your building. If you do have this ability, see more options in the Variations section.)

4) Connect a 3" elbow to the adapter. The exact angle this elbow needs to be will depend on your particular situation. Because 90° elbows can create a lot of turbulence, opt for an angle that's less than 90° if you can. Ideally, use a 3" pipe strap to connect the elbow and adapter to the side of the building in case of high winds.

5) Next, cut a hole into the top of the cistern where the basket strainer will sit. Follow your basket strainer's manufacturer's guidelines on how to measure and cut the right-sized hole on top of the cistern using a holesaw or jigsaw. The basket strainer does not need to be same size as your pipe. For example, you can use an 8" basket strainer. Once the hole is cut, pop the basket strainer into the hole.

6) Using the horizontal and vertical distances you measured from the top of the tank to the bottom of the gutter, calculate or measure how long the inflow pipe will need to be. Cut the pipe to length and sand off any burrs. To prevent mosquitos from getting into the pipe, attach the pantyhose around one end of the pipe and zip-tie it far enough down the pipe that you can still cement the pipes together. Connect the horizontal pipe to the downspout adapter using PVC cement and then place the end of the horizontal pipe on the basket strainer.

7) You may notice that the end of the pipe is open to animals/insects. Install a 3" hinged vent flap on the end of the horizontal pipe (or alternatively, zip-tie on a set of pantyhose) for extra filtration and mosquito prevention.

5. Route the overflow away from the adjacent building

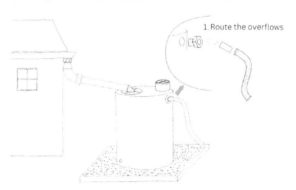

1. Route the overflows

1) Measure the diameter of the top outlet port. This will be used for overflows. For this tank, the outlet is 1.5". Screw a 1.5" male bulkhead adapter into the female bulkhead fitting. Connect a 1.5" elbow to the male bulkhead adapter, followed by a 1.5" straight pipe of your desired length. This pipe can be as long or as short as you need it to be, as the rest of the overflow can be routed away using a flexible pipe like a piece of 1.5" corrugated tubing.

6. Configure the outlet and enable the use of rainwater

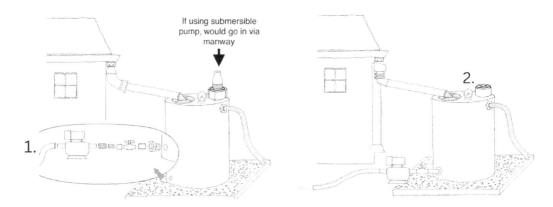

1) Measure the diameter of the bottom outlet port. This is where the hose or pump will connect. For this tank, the outlet is 2". Screw a 2" male bulkhead adapter into the female bulkhead fitting, then attach a 2" ball valve to act as a shutoff. From there, you can configure the outlet to match the size of the outlet you need.

- For a garden hose, connect a 2" to ¾" reducer bushing to the valve, then add a ¾" MHT adapter that will allow you to screw on a garden hose.

- For an external pump, connect a bushing that matches the size of the pump inlet, then attach a nipple. Follow that with piping (such as PVC or PEX) that's the same size as the pump inlet. Follow pipe guidelines for leak-proofing and installation when you do this.

- For a submersible pump, you'll need to use a manway to drop in the submersible pump. Connect the outlet of the submersible pump to the inside of the bulkhead fitting using flexible corrugated tubing in accordance with the size of the pump outlet. Install any required bushings.

2) To install a tank gauge, follow the manufacturer's instructions for how to drill the right-sized hole at the top of the tank and calibrate the tank gauge. Drop the tank gauge weight into the tank and secure the gauge readout to the top of the tank with screws.

Variations:

- If you're using a metal tank, use finely crushed rock or paver rock rather than gravel— the finer rocks won't dent the tank as easily.

- If you have the ability to screw pipe straps into the side of your building (perhaps you would be attaching them to wooden posts or you have the equipment to tap into brick or rock), attach a Leaf Eater by placing a straight pipe onto the side of the building and then placing the Leaf Eater on top of the pipe. Add the elbow at the bottom of this pipe to connect it to the horizontal inflow pipe. This avoids the need for ugly pantyhose while keeping the pipe filtered and animal/insect-free. For greater insect control, consider adding a hinged vent flap (also known as a Mozzie Stoppa) on the end of the downspout that's going into the Leaf Eater.

- In addition to the filtration methods described so far, also consider a floating filter. These can be purchased online and generally offer finer filtration than basket strainers or in-line filters do. Floating filters are handy for allowing finer filtration before the rainwater enters the pump. Once you purchase one, follow the manufacturer's guidelines for installing it and attaching it where you need it to be.

HOW TO CONNECT TWO OR MORE TANKS TOGETHER

You may occasionally see setups for rainwater harvesting systems where there are two or more tanks connected together. This allows the harvester to gradually increase the size of their rainwater storage over time and modulate their storage based on their spatial constraints (among other benefits). Then again, sometimes it can be easier to simply purchase one bigger tank than to attach another tank to one you already have.

Start with one tank first. If you find yourself wanting to connect two or more tanks together, then follow this guide. These instructions use the same two cisterns in the cistern example to illustrate how to join two tanks, but these same instructions can be adapted for rain barrels, IBC totes, etc. Use the correct sizes and fittings for your particular tanks. There are two ways to connect two or more tanks together: near the top of the tanks or near the bottom of the tanks. When you connect two or more tanks together near the top, this is also known as connecting tanks "in series": the first tank fills up, then overflows into the next; after that, the overflow tank starts to fill and then may overflow into another tank, and so on. Essentially, tanks connected near the top or "in series" are connected via their overflow outlets. When you connect two or more tanks together near the bottom, this is also known as connecting tanks "in parallel": when the first tank fills and the water reaches the hole at the bottom, water spills into the pipe that connects the tanks near the bottom, filling the other tank until both tanks have reached the height of the bottom hole. Then the two tanks fill up at the same level. Essentially, tanks connected near the bottom or "in parallel" are connected via their bottom outlets.

TABLE 15: PROS AND CONS OF CONNECTING TANKS AT THE TOP VS. AT THE BOTTOM

	Pros	Cons
Tanks Connected at the Top (In Series)	Fewer tank(s) will fill completely, potentially decreasing maintenance needs	A need to install extra bypass outlets to access water in other tanks
Tanks Connected at the Bottom (In Parallel)	Essentially multiply your storage capacity by the number of tanks connected together	Might result in low water pressure if you're not using a pump

HOW TO CONNECT TWO OR MORE TANKS AT THE TOP (IN SERIES)

Skill level: Intermediate

Approximate time: 60 – 120 minutes

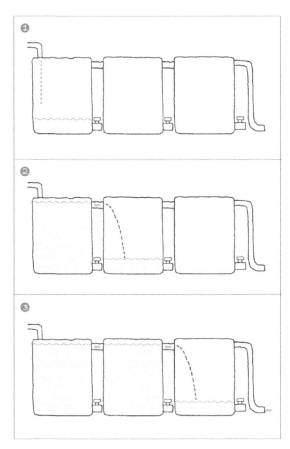

Water overflowing from one tank to the next do so via tanks connected at the top (in series).

Equipment needed:

- PVC cement
- Teflon tape
- Screwdriver
- Hacksaw
- Sandpaper

Materials needed:

For the connection(s); 1 connection between each tank:

- (2) 2" PVC bulkhead adapters
- (2) 2" PVC straight pipes (3" – 5")
- (1) 2" flexible coupler

For the bypass outlets (for all tanks except the last one):

- (1) 2" to ¾" PVC reducer bushing
- (1) ¾" PVC straight pipes (3" – 5")
- (1) ¾" ball valve
- (1) ¾" MIP to ¾" MHT adapter

OVERFLOW OUTLET
CONFIGURATION

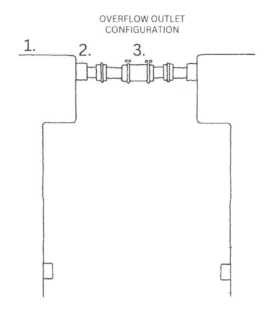

1. Turn the two tanks so that their overflow outlets are facing each other.

2. Make sure to use Teflon tape or PVC cement **between each fitting** to keep all fittings leak-proof. Connect a 2" PVC male bulkhead adapter into each overflow outlet on each tank. After the bulkhead adapter, attach a 2" PVC straight pipe into each tank. Push the two tanks together until they're only a few inches apart. This space will be connected by the flexible coupler.

3. Connect the two tanks together with a 2" flexible coupler and tighten the coupler's screws with a screwdriver. You can use any coupler in this scenario, but a flexible one allows for slight movement in case either of the tanks expand or shift due to changes in temperature or because the foundation settles.

4. Repeat for any other tanks you may have. Test to see how your setup works by filling the tanks with a hose.

To access the water in all of the tanks, you'll need to configure the outlets on all of them. While the final outlet to the hose or pump comes out of the final tank, you may need to have a bypass outlet on any of the tanks in the series in case the water level in the last tank is very low.

To configure the outlets on any of the tanks before the last one, configure the outlets to have valves at the end so that they can be opened or closed depending on your needs.

To configure a bypass outlet on the bottom outlets of all of the tanks:

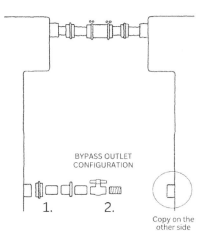

1. Screw a male bulkhead adapter into the bottom female bulkhead outlet. Connect a 3" – 5"-long 2" straight pipe after that, followed by a 2" to ¾" reducer bushing.

2. Connect a 3" – 5"-long ¾" straight pipe to the reducer bushing, followed by a ¾" isolation ball valve. After the ball valve, connect a ¾" MIP to ¾" MHT adapter (or any adapter you need—this is an example that attaches to a garden hose). Keep this valve closed when using water from the last tank, but open it when you need to connect a pipe or hose straight to the outlet in any of the other tanks.

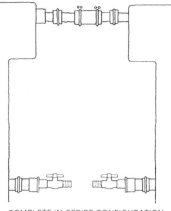

COMPLETE IN-SERIES CONFIGURATION

These steps apply to 3 or more tanks as well—simply apply the steps to as many tanks as you need. To access water in all of the tanks, configure outlets on all of the tanks to bypass the connection in case you want to access water in any one particular tank.

HOW TO CONNECT TWO OR MORE TANKS AT THE BOTTOM (IN PARALLEL)

Skill level: Intermediate

Approximate time: 60 – 120 minutes

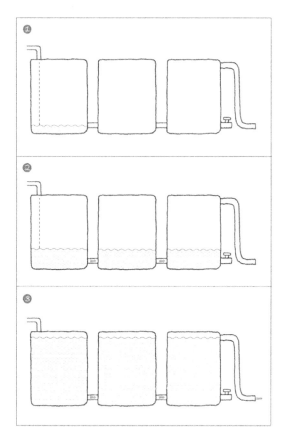

Water flowing from one tank to the next via their bottom outlets do so via tanks connected at the bottom (in parallel).

Equipment needed:

- PVC cement
- Teflon tape
- Screwdriver
- Hacksaw
- Sandpaper

Materials needed:

For the connection(s):

- For the connection(s); 1 connection between each tank:

    - (2) 2" PVC male bulkhead adapters
    - (4) 2" PVC straight pipe (3"– 5")
    - (2) 2" ball valves
    - (1) 2" tee
    - If connecting 3 or more tanks: (1) 2" flexible coupler

- For the outlet:

    - (1) 2" PVC straight pipe (3"– 5")
    - (1) 2" to ¾" reducer bushing
    - (1) ¾" MIP to ¾" MHT adapter nipple

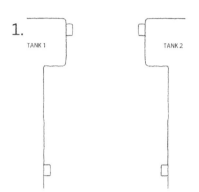

1. Turn the tanks so their bottom outlets are facing each other.

2. Make sure to use Teflon tape or PVC cement **between each fitting** to keep each fitting leak-proof. Connect a 2" PVC male bulkhead adapter into each overflow outlet on each tank. After the bulkhead adapter, attach 2" PVC straight pipe into each tank.

3. After each straight pipe, attach 2" isolation ball valves, then a 2" straight pipe.

4. Push the two tanks together until they're only a few inches apart. The space between the two tanks will be connected by a 2" tee. First, connect one side of the tee to one outlet, then push the two tanks together until the other outlet fits into the other side of the tee. The outlet of the tee should be facing outwards. You'll install two isolation valves here to allow you to perform maintenance on one tank without losing the water in the other tank.

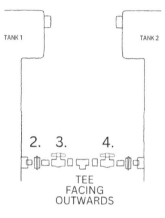

5. Configure the tee outlet so that you can attach a pipe or hose to it. Follow these instructions to attach a hose:

   (1) Connect a 3" – 5"-long 2" straight pipe to the end of the tee, followed by a 2" to ¾" reducer bushing.

   (2) Connect a ¾" MIP to ¾" MHT adapter to the end of the bushing (or use any adapter you need—this is an example that attaches to a garden hose). You can attach another isolation valve or hose bibb to the end of this adapter, or you can attach a garden hose straight to the adapter.

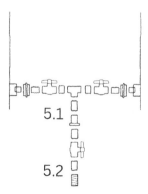

If you have 3 or more tanks...

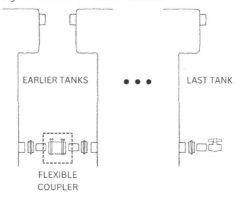

EARLIER TANKS   • • •   LAST TANK

FLEXIBLE
COUPLER

To connect 3 or more tanks in parallel, instead of using a tee, connect the isolation valves between the tanks with a flexible coupler.

HOW TO INSTALL A FIRST-FLUSH SYSTEM

This installation guide can be used and adapted for any of the preceding installation guides. Note that you'll need a first-flush diverter kit, which is available online from Amazon, RainHarvest.com, etc. Aside from the kit, you can find the other materials at a hardware store. To incorporate this plan into any of the other setups, simply attach the horizontal inflow pipe to the tee. The other side of the tee fits into the downspout and the first-flush storage pipe. You can also attach an in-line filter like Leaf Eater to the top of the tee. This is an example of a first-flush system that you can not only use for the IBC tote setup but can adapt for any setup.

Skill level: Intermediate

Approximate time: 1 – 2.5 hours

Equipment needed:
- Measuring tape
- Hacksaw
- Sandpaper
- PVC cement

Materials needed:

- (1) first-flush diverter kit (pick a size that best accommodates your downspout; e.g. 4")

  - Should include:

    - (1) tee
    - (1) ball
    - (1) endcap with socket
    - (2) pipe straps
- (1) downspout adapter (dependent on the size of your downspout)
- (8) downspout screws or sheet metal screws (⅜" – ½")
- (1) in-line filter (e.g., 4" Leaf Eater)
- (1) 4" PVC straight pipe (the length will be determined by your measurements)

1. If you wish to use a first-flush diverter, use these instructions when configuring the downspout and inflow of your particular system.

2. Your first-flush diverter kit should come with a tee, ball, endcap with socket, and possibly pipe straps. You will likely follow your manufacturer's instructions.

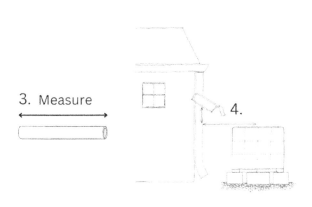

3. Measure

3. This setup assumes that you have not cut or configured your downspout yet; it also assumes that you're using an in-line filter on top of the tee. First, measure the length of the pipe you need for your first-flush diverter. The length will depend on your roof size; for a 2,000 ft² roof, usually 8'-long pipe with a 4" diameter will suffice. Your pipe may need to be shorter depending on how much vertical space you have. Consider configuring the height of your inflow pipes to match the height of your pipe accordingly.

4. The horizontal inflow pipe will fit into the tee. Measure how high up you'll need to place the tee so that the inflow horizontal pipe matches the height where it goes into the IBC tote. You also need to allow the pipe to slope downwards.

5. Cut the downspout to fit this height and sand off any burrs. Most first-flush diverters use a round pipe configuration. If you have a rectangular downspout and/or a pipe that's not the same size as your tee, screw the adapter onto the end of the downspout.

6. Off the wall, connect the first-flush pipe to the tee and the tee to the in-line filter.

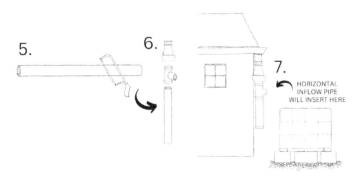

7. Using pipe straps, mount this pipe 6" – 8" below the downspout adapter. If you don't have the ability to tap into the side of the building or don't want to, consider supporting the end of the pipe on the ground by using an extension of the pipe and an elbow.

8. Place the ball inside the pipe on the other end. Screw on the endcap with a socket.

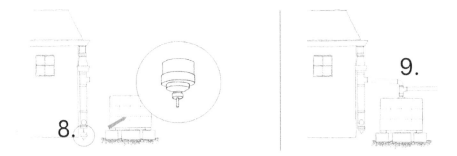

9. Connect the horizontal inflow pipe to the tee.

10. Follow the manufacturer's instructions for winterizing your first-flush system. The ball may crack in cold temperatures, so be sure to keep the system empty and possibly remove the ball before a deep freeze.

HOW TO INSTALL A RAINWATER HARVESTING CHICKEN WATERING SYSTEM

This setup is inspired by Joe at Homesteadonomics, @homesteadonomics. Joe is a resourceful homesteader who lives fully off rainwater in the Arizona desert. This guide routes water from the roof of a chicken coop to a rain barrel and then to a pipe that feeds chicken watering nipples. Therefore, this watering system is helpful for chickens who already know how to peck at watering nipples. (That said, you can train them by placing a sunflower seed in the cup of the nipple—then they'll peck to activate the cups.)

This setup installs the chicken watering pipe inside the coop where the chickens usually drink, which is usually in the run. Your particular coop layout may look different.

Skill level: Intermediate

Approximate time: 2 – 4 hours

Equipment needed:

- Electric drill
- ³⁄₈" brad point bit
- PVC cement

- Silicone caulk
- Teflon tape
- Pliers/crimpers

Materials needed:

For the water pipe:
- ¾" PVC straight pipe (the length will be determined by your measurements)
- ¾" PVC threaded endcap
- ¾" PVC male MHT adapter
- (3 – 4) ¾" pipe straps

For the roof/gutters:
- Roof flashing or aluminum gutters (the size is up to you; it can be around 3" – 4" round style)

For the rain barrel:
- (2) cinder blocks
- (1) prefabricated rain barrel with an inlet screen, overflow outlet, and outlet hose threads already installed (see How to Install a Rain Barrel)
- (1) flexible pipe (the same size as the outlet on the rain barrel)

For the downspout and inflow:

- (1) ¾" 90° elbow
- (1) ¾" PVC straight pipe (the length will be determined by your measurements)
- (1) ¾" stainless-steel hose with ¾" female hose threads on both ends

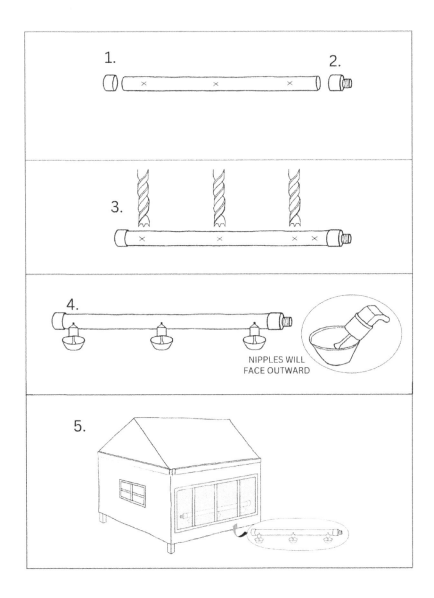

1. First, build the pipe for the watering nipples. Measure how long you expect your pipe to be. It can be any length, but place each nipple at least 3" away from the next one. Also, a rule of thumb is to have at least 1 nipple per 9 hens.[3] Cut the pipe to length and sand off any burrs.

2. On one end of the pipe, attach a ¾" PVC endcap; on the other, attach a ¾" male adapter with a ¾" male hose thread. Attach all PVC pipes with PVC cement.

3. Drill holes into the pipe with the brad point bit, drilling about 3"– 8" apart.

4. Screw a chicken watering nipple into each hole, making sure all cups face the same way. Add silicone caulk around the edges of the holes to prevent leaks.

5. Install the water pipe somewhere inside the coop, such as in the coop's run. Use ¾" pipe straps to mount the pipe to the beams inside the coop. Be aware that the threaded end should face where you plan to place the barrel.

For the rain barrel:

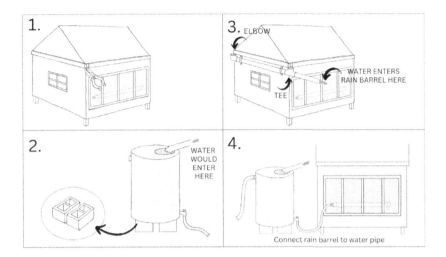

1. If you don't have gutters on your chicken coop, you can add gutters, or you can add some aluminum flashing off the coop's roof using some sheet metal screws that go along the length of your chicken coop's roof. But beware! The flashing won't make for super large gutters, and they may overflow. If you get heavy thunderstorms, you may need full gutters. If you decide to use flashing, use pliers or crimpers to fashion a makeshift downspout at the ends of the roof. If you're using gutters, follow your manufacturer's guide to install a 6" – 8" downspout at the ends of the roof.

2. Place cinder blocks or an elevated platform of your choice near the water pipe, positioning them on the side of the water pipe that has the threaded male end. Because this is a gravity-fed system, the rain barrel will benefit from being elevated.

3. Connect the flashing or gutters on the coop's roof to the rain barrel. Do this by installing the horizontal ¾" PVC pipe (determine the length by measuring the distance from the downspout to the rain barrel) and keeping the pipe attached to the coop using pipe straps. Maintain an 1/8" per foot downward slope on the horizontal pipe. Connect the downspout to the horizontal pipe using a ¾" 90° PVC elbow. Collect rain from both sides of the roof by installing flashing or gutters on the other side of the roof and installing a ¾" tee in the middle of the inflow pipe right where the downspout of the flashing or gutters is.

4. Configure the overflow on the rain barrel by connecting flexible piping to the rain barrel's outlet and directing the water away from the coop (possibly to a nearby garden bed).

Connect the rain barrel to the water pipe:

1. All threaded connections should have Teflon tape in between the threads. Connect one end of the ¾" stainless-steel hose to the outlet of the rain barrel. This assumes that the end of the rain barrel has male hose threads. If the outlet isn't configured for female hose threads, follow the rain barrel installation guide for how to do this.

2. Connect the other end of the ¾" stainless-steel hose to the ¾" male hose thread adapter on the water pipe inside the coop. You may need to cut the screen or wire of the chicken coop very slightly in order for the pipe to fit, or you might need to move the chicken wire out of the way of the hose.

3. Because this is a gravity-fed system, water will enter the water pipe as long as the water level in the rain barrel is at the same level as the outlet. You can manually fill the rain barrel if the water level gets too low.

Congratulations! You're done!

Variations:

- If you need greater water pressure than just what gravity can provide, consider installing a small pump between the barrel and the water pipe. Carolina Coops, a firm that builds beautiful customized chicken coops, offers full kits online with the pump and rain barrel included. This means you don't necessarily need to source the items on an individual basis.

- Instead of using cup-like watering nipples, use watering nipples without the cup (they are often less expensive).

HOW TO CONNECT A RAIN STORAGE TANK TO INDOOR APPLIANCES

This guide will illustrate how to bring rain that has come from the roof and gutters and then been coarsely filtered into the storage tank to where you may need it the most: your indoor appliances. You'll see what's needed to make rainwater suitable for indoor non-potable and potable use. Indoor potable use filtration and treatment is not much more complicated than it is for indoor non-potable use—for example, you just add on another filtration cartridge and treatment option.

Interestingly, the information that many people need to know the most specifics about is actually the most difficult scenario to provide specific examples for. That's because the specifics and variations of an individual setup vary so widely and can be tailored very specifically to meet each person's/household's needs. Given that wide range of possibilities, this guide will provide an illustration of a generic setup as well as an example of a specific setup (the latter uses an external pump with an attached pressure tank as well as a UV disinfection treatment system). You can adapt this guide to your and your family's needs or choice of filtration or treatment. If you have granular plumbing and/or electrical needs and questions, consider visiting your hardware or plumbing supply store to ask these specific questions or consult a plumber or an electrical technician. These experts can provide more specific answers for your particular situation. If you're off-grid, make sure your electrical system is configured to meet the voltage and amperage demands of any of the pumps or treatment options you purchase. (You may need inverters, rectifiers, etc.)

You can use this guide to adapt your plumbing to flush your toilet with rainwater, run your washing machine with rainwater, and even drink and cook with rainwater. This installation uses a setup where non-potable water feeds non-potable water needs (such as the toilet and washing machine) and potable water feeds potable water needs (such as a kitchen faucet).

Skill level: Advanced

Approximate time: 6 – 10+ hours (depends on your level of preparedness and the adaptability of your indoor plumbing)

Equipment needed:

- PVC cement/all-purpose plumbing cement
- Teflon tape
- Wrenches
- Pipe cutters
- Screwdrivers

Materials needed (the exact number will vary depending on your particular needs and setup):

- Ball valves
- Elbows, tees, straight pipes
- (1) pump
- (1) pressure tank (if your pump doesn't already come with a pressure tank)
- Check valve that is the same size as the pump ports
- Filtration

- 100-micron filter
- 20-micron filter
- 5-micron filter

- Treatment

- UV lightbulb and case with control panel

First, let's look at a general diagram of how the storage tank is connected to the pump, then to filtration, then to treatment, and then to your appliances.

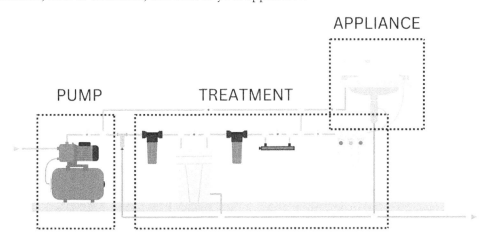

Next, let's look at how this diagram could be adapted for an external pump with a pressure tank and a UV light treatment system.

General guidelines:

- Use isolation ball valves on either side of pumps, pressure tanks, filtration components, and treatment components in order to make it easier to change out components or perform maintenance tasks. Install pressure gauges after the pump so that you can keep an eye on the pressure of the water coming out of the pump.

- The recommended pumps in this book should generally only be used for clear water. If you find you still have a lot of sediment in your tank as the water leaves the tank and enters the pump, that could damage the pump. Consider adding a floating filter to your tank or adding a 100-micron filter before the pump.

- Connecting an appliance to a pump generally requires connecting a supply line with a supply stop valve to a main line that runs throughout the house. The pump then feeds into the main line. For main lines in the house, use 1" pipes—then you can reduce that size to ¾" pipes or even smaller for other lines coming off the main line to the appliances. **Do not use PVC for hot water.**

- If using UV light treatment, do not use PVC or CPVC pipes for the pipes leading into the UV light component, as UV will degrade the components. Use copper or stainless-steel pipes instead.

- UV light treatment can only work on water that has been passed through a 5-micron or finer filter—any bigger, and bacteria can hide behind too-big particles and never get zapped by the UV. If you're planning on using a UV light treatment system, install a 5-micron or so filter right before it.

- If you're connecting a rain tank to the inside of your home as a replacement for municipal water, you may need to build a new main line from the rain tank to the inside of your home. Consult a plumbing professional for particular steps on how to do this.

- If you're building a home or building a line from the rain tank to the inside of your home, you'll simply need to follow best plumbing practices for selecting and building a water supply line for the home. Use 1" PEX, CPVC, copper, or stainless-steel pipes for inside the home and double-check that your materials are rated for the temperature of the water it will be used at.

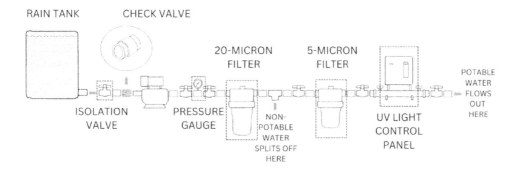

1. This setup will house the pump, filtration, and treatment system on a panel inside the home (such as in your basement). On your storage tank, connect the right-sized adapters to a hose or pipe that has the same diameter as the inlet port on your pump.

2. Before connecting the hose or pipe to the pump, connect an isolation ball valve. This connects to a check valve, which is connected right into the inlet port on the pump. The pump will need to be powered and thus be close enough to the correct power source.

3. On the pump, connect an isolation ball valve, followed by a pressure gauge, followed by another isolation ball valve.

4. To that isolation ball valve, connect the indoor non-potable filtration device (such as a 20-micron filter). Since this setup separates non-potable indoor water from potable indoor water, a tee is then connected after the 20-micron filter to bring non-potable water to toilets, washing machines, etc. See the Variations section for other versions of this.

5. Following the 20-micron filter, connect another isolation ball valve, followed by a 5-micron filter, followed by another isolation ball valve.

6. Following the filters, connect the UV light to the ball valve (make sure that the UV light is connected to the correct power supply). After the UV light, connect the main line.

7. Following the main line, connect the appropriate bushings and stop valves to connect the non-potable supply line to the toilet and washing machine, and then the potable supply line to the faucet. If you're disconnecting the supply lines from the municipal water, you'll need to connect the supply lines to a new main line coming from the rain tank. Follow plumbing best practices to ensure that the main line is connected correctly. When in doubt, consult plumbing experts!

Variations:

- More water is used for non-potable applications than potable applications around the home and garden. Therefore, your entire setup could include two different tank systems, where one connects to non-potable uses and the other connects to potable uses. If you were to connect all of the water inside your home to the potable main line, this would simplify the installation. However, this setup may end up being more costly over the long run due to the need to treat that much water for potability.

- Search online for filtration and treatment kits that have all of the filtration and treatment components in one easy-to-install panel that can be quickly mounted and connected to water supply lines.

ACTION GUIDE FOR CHAPTER 6: INSTALL YOUR SYSTEM

Using your components and keeping the basics of safety and use in mind, install your system using one of the installation guides in this chapter. Alternatively, you could use a combination of techniques from the different types of filtration systems, piping, tanks, and foundations. Be creative and sketch out your system first, considering where you'll position your tank(s) and how your gutters and downspouts will bring rainwater to your tank(s). Have fun!

CHAPTER 7

MAINTAIN YOUR SYSTEM

T o keep your system in tip-top shape, you'll need to maintain it. No problem! Just follow these guidelines. Every time it rains, clean out all of the filters and screens on the downspouts and clear the tank of leaves and twigs. As you clean, inspect the screens for any damage and repair or replace them as necessary.

Leaves and other organic matter will accumulate in your screens. Simply scrape them off.

If you have a first-flush diverter, cleaning out the bottom of the endcap after every rain event or every other rain event will keep bacteria from colonizing there. If you have a version that doesn't self-empty, you'll need to set reminders to empty it out after every rain event. Perform routine visual inspections on the most finicky components of your system, such as pumps, filtration systems, and first-flush diverters. Do this as often as weekly during the rainy season. (These inspections are more commonly performed on a biweekly or monthly basis.) Check that everything is working as it should be and make sure there are no leaks. Before the rainy reason, clean off the roof and gutters and trim back any overhanging branches or foliage.

Clean off your roof and trim back branches frequently. Follow safety guidelines when on your roof!

If the surrounding trees or vines above your roof tend to drop leaves or other debris during a particular time of the year, do more frequent maintenance during that time period. Keeping the roof and gutters clean keeps the rest of the system cleaner and also keeps maintenance needs lower. But regardless, gutters should be cleaned at least once or twice per year; during the rainy season, they should be cleaned once a month to ensure an adequate flow of water from the roof, through the gutters, and into the storage tank. If you use a sand or gravel base for your tank, inspect it monthly to see if any sand or gravel has been washed away by rain and needs to be replaced. You can perform visual inspections of the rain tank, the pipes that come from the downspout to the rain tank, the base, and other non-moving parts less frequently, around once a month (or less often). Open the outlet at the bottom of the rain tank and ensure that water is still flowing through. Check to make sure the vent and overflow outlets are not clogged. Check to see if all of the pipes are still connected and if they may need to be replaced. If you're using PVC pipes, inspect them for any discoloration and/or warping of the PVC. If any of that occurs, replace the PVC pipes. PVC degradation can be slowed by painting the PVC opaque or by using thicker PVC pipes (such as Schedule 80 pipes, which are a thicker version of PVC than the more common Schedule 40 pipes).[1] You can visually inspect the inside of the rain tank for animals or mosquitos as often as weekly, but as long as the screens on the tank are intact, this inspection can be performed on a monthly basis (or less often).

DE-SLUDGING

Over time, very fine sediment passing through your downspout filters will end up at the bottom of the tank, where it will form a layer of organic biomass called the "sludge layer." The sludge layer should be cleaned from the inside of the tank at least every two years,[2] as sludge layer buildup can prevent water from flowing freely. The sludge layer can be removed by siphoning it out or by emptying out the tank completely and then rinsing the inside of the tank. Sludge removal can be easily outsourced to a professional landscaper (or anyone looking for an odd job). But don't clean out the sludge layer too frequently, especially if you don't use a first-flush diverter—the sludge layer plays a helpful role in keeping heavy metals and other toxins out of the water. Removing the sludge layer every two years is sufficient.

ENCLOSED SPACES

There may be times when your rain tank needs servicing from the inside (such as when you're cleaning it or replacing nuts). Your cistern should have a manway or utility hole to allow you to access its interior during maintenance and interior servicing. But before you enter, have an emergency exit plan! Confined spaces lack oxygen, have poor air quality, and can present physical hazards. Work with a buddy (or several) and don't rely on a cell phone or any technology that could malfunction in water to communicate with the outside world. If you're unsure of your ability to do interior tank servicing, consider hiring a professional.

DISINFECTION

If you suspect the water in the tank is contaminated with disease-causing microbes, you can chlorinate it. It's best to use liquid sodium hypochlorite (also known as Pool Shock or "liquid chlorine" in the swimming pool world) or granulated calcium hypochlorite (also known as "granular pool shock").

These can be bought online or at your local hardware or pool store. If you're using liquid sodium hypochlorite, use 1.5 fluid ounces for every 260 gallons of water. If you're using granulated calcium hypochlorite, use 0.25 ounces for every 260 gallons of water.[2] The water may smell and/or taste like chlorine for a few days, but it's safe to drink.

If the roof and gutters are kept clean and the filters are working, the risk of contamination is low. Regular disinfection of the tank is not necessary, as the chemicals used to disinfect the tank (like liquid chlorine) may end up doing more harm than good if used too frequently.[2] Use filtration and potable water treatment for potable water rather than frequent disinfection of the tank.

If you do decide to perform a lab test on the water, instead of testing water from the tank, take water samples at the point of use (i.e., from the end of the hose or from the faucet). It's best *not* to test the water quality from the tank, as lab results can swing in directions that can just make you paranoid. If you're testing for *E.coli*, take a sample that's a minimum of 3.5 fluid ounces.[2] Send results to a nearby water testing facility to receive results. (Search online for "water quality testing near me" or "well water quality testing near me.") Your municipal or county health department can also provide a list of reputable water testing facilities in your area. If *E. coli* is detected, then you should chlorinate the tank. First add chlorine to the water, then test the results again. If you continue to detect contamination, you may need to dump out the contents of the tank and disinfect the inside of it. You can schedule your water quality testing on a quarterly basis.

PUMP MAINTENANCE

You'll perform routine pump maintenance by following your pump manufacturer's guidelines, so don't throw that manual away! Those instructions will let you know how frequently you need to perform maintenance tasks and what you need to do (e.g., which lines need to be disassembled, which parts need to be inspected, what number to call if your pump needs servicing, etc.). If you need to unhook a pump from any electrical or plumbing connections, schedule that downtime in advance— preferably have it be when you won't need the pump for a little while.

A residential property using an external pump and pressure tank for its rainwater tank.

When your pump arrives and you first install it, get to know your pump and observe how it sounds, looks, and smells when it runs. That way, if there are any leaks or strange sounds or smells, you'll be familiar enough with the pump to know if the pump needs servicing or if you need to call your manufacturer. If you aren't confident about servicing your pump yourself, have a professional service it at least the first few times. Then you can learn how they diagnose and service pumps. If you have a submersible pump, you may need a buddy or a professional to help you climb or reach down into the tank, pull out the pump, and look at it.

Regularly check to make sure that your pump's mounting points are still secure and haven't vibrated off. Also check for any leaks and make sure the pump still seems to be sealed. Look for any fraying of electrical wires and inspect and clean any filters that may have come with your pump. If you have a smaller pump, check the pump for any leaves or debris that may have passed through your initial filtration system. That debris can clog your pump and prevent it from working efficiently, which can drive up your costs to run the pump. How often you'll need to do all of these tasks will depend on your pump, but you'll likely be doing this maintenance on a quarterly or semiannual basis.

WINTERIZATION

You'll need to winterize your system when temperatures start to dip below freezing in your location. Dumping out the tank is not always necessary, especially not if you live in a climate that doesn't get extremely cold, but do follow these steps to keep your system functional for the next spring:

1. If you have a rain barrel, empty it out. For tanks larger than a barrel, keep the tanks as full as possible—that will make it less likely that the tank will freeze entirely. However, if you live in an area that gets colder than Zone 6b (search online for "USDA plant hardiness zones" to find out what your zone is) or if you live higher than 3,000 feet above sea level, you'll likely need to empty out the tank enough so that ice expansion won't damage the tank. How much you'll need to empty out depends on your climate and how cold your climate gets. If in doubt, consider emptying out the entire tank.[3]

2. If you have a submersible pump, it might be better to remove it entirely, but if you live in climates like the southeastern regions of the US where it doesn't get very cold for very long, you might be able to get away with leaving the pump in the tank. **Only do this with a full cistern of water.** You can only do this if the risk of the full cistern of water freezing is low to none, thereby lending little risk to damaging the submersible pump. If you have an external pump, you'll need to drain the pump and store it inside your home or in a heated area during the winter.

3. Double-check the inlet pipes going to the storage tank to make sure they're clear of standing water. Inlet pipes can still be routed to the tank if you live in Zone 8a or warmer, but otherwise, inlet pipes need to be routed away from the tank so that they won't continue to fill the tank during the winter months.

4. Open spigots slightly, just enough so they drip. Hoses and drip irrigation systems need to be drained and brought inside. Also bring inside any electronics you're using, such as timers and especially pumps.

Consider the annual rhythms of an aboveground rainwater harvesting system: commission it for the spring and decommission it for the winter, the same way you would treat a garden.

Rain barrels in a freezing climate. If using a downspout diverter, ensure that the diverter doesn't divert water to the rain barrels during the winter to maintain your winterization efforts.

RAINWATER HARVESTING SYSTEM MAINTENANCE SCHEDULE

The best way to prevent the tank and the water inside it from getting contaminated is to keep the roof, gutters, screens, and filters leading into the tank as clean as possible. It's also crucial to keep the tank sealed from insects and animals. This level of "sensible maintenance," as you could call it, has served 6.3 million Australians *without* including frequent disinfection and chlorination and without any reported widespread outbreaks of disease so far. See Table 16 to follow a guided schedule of maintenance. You can adopt this schedule for your own needs or adapt it as your specific situation necessitates.

TABLE 16. RAINWATER STORAGE TANK MAINTENANCE SCHEDULE

Every time it rains/ every other time it rains	Weekly	Monthly	Quarterly	Annually	Every 2 Years	As-Needed/ Ad Hoc Basis
Clean/inspect the filters and screens	Visual inspections of pumps and indoor-use filters	During the rainy season, quickly clean the gutters to keep water unobstructed and clean	Test water quality for heavy metals and/or bacteria	Before the rainy or foliage fall season, clean off the roof	De-sludge the tank	Test water for heavy metals and/or bacteria if you have reason to suspect contamination
Clean out first-flush diverter		Inspect the gravel or sand base to see if any material needs replacing	Do routine pump inspections and maintenance	Semiannually to annually, clean out the gutters		
		Visually inspect the pipes, tank (inside and outside), outlets, overflows, and vents		Winterize the system		

ACTION GUIDE FOR STEP 7: MAINTAIN YOUR SYSTEM

Step 1:

Once your system is installed, follow the maintenance schedule table in this section to establish your family's schedule for your very own rainwater harvesting system.

Step 2:

When temperatures start to dip below freezing, plan to winterize your system and create a schedule to do so. Depending on how cold it gets, your particular climate may allow you to continue to use your aboveground system during the winter. Keep your cisterns as full as possible, and if you'll need to decommission your rainwater harvesting system during the winter months, arrange for secondary sources of water to be brought in during that time (e.g., well water, trucked-in water).

Leave a 1-click review!

Customer reviews

★★★★★ 5 out of 5

20 global ratings

5 star		100%
4 star		0%
3 star		0%
2 star		0%
1 star		0%

˅ How are ratings calculated?

Review this product

Share your thoughts with other customers

Write a customer review

To leave your review on Amazon, just visit:

www.review.reneedang.com

Or, scan here with your phone!

If you enjoyed this book, I would be very grateful if you could take just 60 seconds to leave me your review of this book on Amazon.

Books like these help other homesteaders and gardeners just like you become more empowered and self-sufficient.

Thank you for spreading the word and supporting indie authors everywhere! Contact the author at renee@reneedang.com.

CONCLUSION

C ongratulations for choosing to harvest rainwater! You are part of a small, tight-knit group that is revitalizing a back-to-roots approach to our time, energy, and resources. Having knowledge of a valuable skill like rainwater harvesting can help you find independence and self-sufficiency almost anywhere you go.

In the US, municipal water can cost as much as $200/month per household, with the costs of watering plants and animals, using sprinklers, and using water for other outdoor applications adding as much as $600/month to a household's water bill. These costs can climb even higher in areas with arid climates, as well as during the summer months when watering needs are higher. Consider this for a moment: can you afford to *not* pick up a rain barrel and catch the free rain already falling onto your yard and property?

Many local governments yearn for homeowners to take their water resources into their own hands so much that the municipality will pay *you* to seek independence from an already-strained grid. Just as cities across the US pay residents to use solar panels to add electricity back into the grid, many cities pay residents to harvest their own rain and keep it out of the municipality's stormwater system. As you probably already know, it pays to be self-sufficient!

A growing population and now climate change are straining water sources that we once took for granted. In 2021, the federal government declared the once-mighty Colorado River—the river that supplies water for an estimated 25 million people across Arizona, Nevada, California, and some of Mexico—to be at a shortage for the first time in history. While rainwater harvesting is not the only solution to a growing water crisis, millions of people around the world are turning back to it as an important strategy to fight water stress. In India, where 54% of the world's second-most-populated country faces high to extremely high water stress, more than half of the states in India *mandate* at least some version of rainwater harvesting. This focus on collecting and using a free and renewable resource in India may reflect what the US will face in just a few short years. As a rainwater harvester, you may be already ahead of the game!

When you begin to harvest rainwater, you'll consciously take stock of your water consumption as an individual and as a household. This revelation will put wasteful practices into perspective and serve as a learning process—you'll actively conserve water and share your newfound knowledge with friends and family as your water bills decrease, your peace of mind increases, and you make greater and greater contributions to your sustainability and self-sufficiency.

Pretty soon, other people will start to turn to *you* to learn how to harvest rain. Spread the message, spread the rain, and happy harvesting!

A free gift to our readers

Learn the 9 Mistakes Contaminating Your
Rainwater Right Now (and how to fix them!)
Grab your free gift by visiting:

www.reneedang.com

Or, scan here with
your phone!

RAIN HARVESTING GLOSSARY

Acid rain: Rain considered more acidic than usual; caused by industrial activity

Berm: A raised mound that keeps water contained within a certain area

Catchment surface: The large surface area over which rain falls; the rain is then diverted to conveyance systems and into a storage tank

Chloramine: Disinfectant used to treat drinking water; formed when ammonia is added to chlorine

Chlorine: An element used for many household applications; colloquially, used as a substitute term for chloramine when chloramine is used to disinfect drinking water

Cinder block: A lightweight building brick made from small cinders mixed with sand and cement

Cistern: A larger tank used to hold rainwater; a cistern typically holds 500 gallons of water or more

Conduit: A channel, tube, or pipe used to transport or carry fluid such as water

Conveyance: See "conduit"

Countertop filtration system: A filtration system that sits on a table or counter that is used to filter and treat water for drinking; non-potable water is poured in at the top and potable water is accessed through a spigot at the bottom

CPVC: Short for chlorinated polyvinyl chloride; a type of pipe material that is similar to PVC but can be used for hot-water applications in the home

Crushed rock: Rock fragments and pebbles created as a byproduct when larger pieces of rock are crushed; typically sold in bags

Downspout: A vertical section of the gutter used to bring rainwater from horizontal gutters down to the ground or into a rainwater storage tank

E.coli: Short for *Escherichia coli*, this is a bacteria spread through contaminated water that can cause stomach cramps and diarrhea, among other symptoms

Earthwork: Earth that is shaped to move rainwater along the ground to follow the lay of the land

First-flush diverter (or first-flush system): A pipe that intercepts an initial finite volume of rainwater runoff from a roof, thereby keeping the dirtiest water out of the storage tank

French drain: A long, narrow, underground trench attached to a pipe; used to divert groundwater

Giardia: Short for *Giardia intestinalis*, which is a parasite spread through contaminated water that can cause diarrhea, among other symptoms

Gravel: A loose aggregation of small, water-worn or pounded stones; typically sold in bags

Gutter: A shallow, horizontal trough fixed beneath the edge of a roof that carries rainwater from the roof to the downspout

Gutter guard: A screen or filter placed on top of gutters that allows rainwater to pass through it but keeps larger debris from entering the gutter

Gutter screen: See "gutter guard"

Hard water: Opposite of soft water; water that has a high mineral content and is hard to sud soap with

HDPE: Short for high-density polyethylene; a type of tank and pipe material that is similar to PVC but can be used for hot-water applications in the home

IBC tote: Short for "intermediate bulk container," these are used for the industrial transportation of large volumes of liquid; IBC totes are used often as inexpensive and readily available rainwater storage tanks and typically hold between 250 to 300 gallons of water

In parallel: Tanks arranged so that water passes through each storage tank at the same time or via the same connection

In series: Tanks arranged so that water passes through each storage tank successively

Micron: A unit of length in the metric system equal to 1 millionth of a meter; often used to measure the size of a filter's holes

Outlet: In regards to rainwater storage, an outlet is a conduit for water to exit a storage tank

Ozone: A colorless, unstable gas with powerful oxidizing properties, used to kill bacteria to render them harmless

Patio paver: A paving stone, tile, brick, or brick-like piece of concrete commonly used as exterior flooring for patios, walkways, driveways, and other outdoor platforms

Potable: Safe to drink; drinkable

PVC: Short for polyvinyl chloride, this is a type of pipe material that is rigid, chemically resistant, easy to install, and inexpensive; cannot be used for hot water

Rain barrel: A barrel used to collect rainwater; typically holds between 50 to 80 gallons of water

Schedule: In plumbing, this refers to the thickness of pipes (e.g., Schedule 40 is less thick than Schedule 80)

Sludge: A layer of organic biomass consisting of sediment and heavy metals that forms at the bottom of a storage tank; considered nonharmful unless disturbed and released into rainwater

Soft water: Opposite of hard water; water that has a low mineral content and is easy to sud soap with

Storage tank: Where rain is held when it falls from the catchment surface to the gutters and into the tank; a vessel for rainwater

Storage tank foundation: A hard, compacted surface that distributes the weight of the rainwater storage tank and perhaps elevates the storage tank

Swale: A shallow ditch with sloping sides that intercepts water runoff

Ultraviolet (UV): A frequency of light too high for the human eye to see; when used to radiate water, UV renders bacteria harmless and unable to reproduce

Vent: In regards to rainwater storage, a vent is a conduit that allows air to pass freely in and out of a tank

Winterization: Preparing a rain storage tank for the winter; in less wintry climates, winterization may mean keeping the storage tank full to avoid full freezing, while in more wintry climates, winterization may mean draining the storage tank completely; the process typically always involves decommissioning the rainwater harvesting system unless the system is insulated or properly heated

PUMP GLOSSARY

Automatic pump: Automatically pumps fluid based on a control (the control usually maintains a certain pressure or flow rate); an automatic pump can also be timed

Booster pump: Refers to a pump used to increase the pressure of a fluid

Cavitation: The formation of bubbles in a liquid; in a pump, this causes pitting on the impeller

Centrifugal pump: Refers to the mechanism that a pump uses to *pump* water; to *move* fluid, a pump uses an impeller

Deep well pump: Refers to a pump used to pump wells more than 25 feet deep; the opposite of a shallow well pump

Diaphragm pump: Refers to the mechanism that a pump uses to pump water; a pump uses a flexible diaphragm and suitable valves to pump fluid

External pump: A pump that is positioned externally to the fluid that it pumps

Impeller: The rotating part of a centrifugal pump; an impeller moves fluid by rotating

Jet pump: Refers to the pressure created by a pump (it creates large amounts of pressure like a jet engine does); a jet pump can be used to draw water from a well

Manual pump: This kind of pump needs to be turned on when you need to use it and then off when you don't need to use it

Pressure tank: A pressure vessel used to hold water at a higher pressure than ambient pressure; used to maintain water pressure as water leaves a pump

Priming: Regarding pumps, priming means pouring fluid into the pump to remove air in the suction line of the pump

Self-priming: Regarding pumps, self-priming pumps are a type of pump that does not need manual priming

Shallow well pump: Refers to a pump used to pump wells 25 feet deep or less; the opposite of a deep well pump

Submersible pump: A pump that's entirely submerged in the fluid that it's pumping

Transfer pump: Refers to a pump used to move fluid from one location to another; also known as a "utility pump"

Utility pump: Refers to a pump used to move fluid from one location to another; also known as a "transfer pump"

PLUMBING GLOSSARY

Ball valve: A type of valve that uses a perforated ball to stop and start water flow; can be used as an isolation valve or other types of valves due to the speed of its closure

Bulkhead fitting: A fitting designed to allow a tank to drain through a hole; creates a seal to avoid leakage

Bushing: An interface between two parts; as a reducer bushing, a bushing allows smaller fittings to fit into bigger ports

Cam and groove: A fitting or coupling used to connect and disconnect two hoses quickly using a male groove adapter with two handles and a female coupler (also known as a "camlock fitting")

Camlock: See "cam and groove"

Check valve: A valve that closes to prevent the backward flow of fluid; used in line within the system

Conduit: A channel, tube, or pipe used to transport or carry fluid such as water

Conveyance: See "conduit"

Copper: A type of pipe material that is rigid, corrosion-resistant, and biostatic; can be used for hot-water applications inside the home

Coupler: A very short length of pipe with a socket at one or both ends that allows two pipes or tubes to be joined

CPVC: Short for chlorinated polyvinyl chloride; a type of pipe material that's similar to PVC but that can be used for hot-water applications inside the home

Elbow: A pipe fitting installed between two pipes that allows for a change in direction

Female: In plumbing, a pipe where the threads are on the inside; screws into male fittings

FHT: Short for "female hose threads," this is the female end of a pipe with hose threads; cannot be used with pipe threads

FIP: Short for "female iron pipe," this is a female fitting with NPT threads; see "NPT"

Foot valve: A one-way valve used at the base of a suction pump; analogous to a check valve used to suction fluid in a vertical direction

FPT: Short for "female pipe threads," this is a female end of a pipe with pipe threads; cannot be used with hose threads; see "NPT"

Gasket: A ring of rubber or other material used to seal a junction between two surfaces

Gate valve: A type of valve that uses a linear-motion gate to stop and start water flow; can be used as an isolation valve when frequent shutoffs are not required

Holesaw: A tool for making circular holes that fits on the end of an electric drill; consists of a metal cylinder with a toothed edge

Hose bibb: An outdoor faucet specifically designed to accept a hose attachment; usually seen on outdoor spigots (or spickets)

Isolation valve: A valve used to stop the flow of water to a particular location, usually for maintenance or safety purposes

Male: In plumbing, a pipe where the threads are on the outside; screws into female fittings

Male adapter: Fits into a female threaded pipe on the other end; connects two pieces of pipe that are the same diameter

MHT: Short for "male hose threads," this is a male end of a pipe with hose threads; cannot be used with pipe threads

MIP: A male fitting with NPT threads; see "NPT"

MPT: Short for "male pipe threads," this is a male end of a pipe with pipe threads; cannot be used with hose threads; see "NPT"

Nipple, hex nipple: A very short length of pipe with two ends, both with male pipe threads

NPT: Short for "national pipe tapered threads," this is a US standard for measuring tapered threads for threaded pipes and fittings; sometimes referred to as "MPT"

PEX: Short for "cross-linked polyethylene," this is a type of pipe material that's used as a flexible pipe for domestic water transport; can be used for hot-water applications inside the home

Plumbing cement: A chemical solvent used to fuse pieces of pipe together; when you're using plumbing cement, you'll need to choose the correct cement for the type of pipes you're using (i.e., what material they're made of)

Plumbing tape: See "Teflon tape"

PVC: Short for polyvinyl chloride, this is a type of pipe material that's rigid, chemically resistant, easy to install, and inexpensive; cannot be used for hot water

Sanitary tee: A plumbing fitting meant to drain water and vent the system; helps connect a horizontal pipe to a vertical pipe

Tee: A pipe fitting with three ports, all connected at 90° in a T shape

Teflon tape: A thin, stretchable tape that is wrapped around threads before threaded pipes are connected to prevent leaks

Union: A type of fitting equipment designed to unite two pipes that can be detached from the two pipes easily without deforming them

Y or wye: A plumbing fitting meant to divert the flow of water; in a downspout, a Y is used to divert water from one side of the Y to another

REFERENCES

Introduction

[1]https://www.harvesth2o.com/incentives.shtml#pa

Chapter 1

[1]https://www.sahealth.sa.gov.au/wps/wcm/connect/public+content/sa+health+internet/public+health/water+quality/rainwater/rainwater

[2]https://www3.epa.gov/acidrain/education/site_students/phscale.html#:~:text=Typical%20acid%20rain%20has%20a,acidity%20is%2010%20times%20greater.

[3]https://hillcountryalliance.org/rainwaterrevival/

Chapter 3

[1]http://www.rcdsantacruz.org/rainwater-harvesting-indoor-uses

Chapter 4

[1]https://www.homedepot.com/p/DRYLOK-Original-1-gal-White-Flat-Latex-Interior-Exterior-Basement-and-Masonry-Waterproofer-27513/100118662?source=shoppingads&locale=en-US&&mtc=Shopping-B-F_D24-G-D24-024_012_WATERPROOFER-Multi-NA-Feed-SMART-NA-NA-FY21_Exterior_SMART&cm_mmc=Shopping-B-F_D24-G-D24-024_012_WATERPROOFER-Multi-NA-Feed-SMART-NA-NA-FY21_Exterior_SMART-71700000081717307-58700006942061315-92700062508088588&gclid=CjwKCAiA78aNBhAlEiwA7B76p4P4LEqCCXiIUN5OQMoBC5HzaYoP6j81eDR7h45eKofxEGKxPpayhBoCOlsQAvD_BwE&gclsrc=aw.ds

[2]Avis and Avis

[3]https://sonomamg.ucanr.edu/files/185639.pdf

Chapter 5

[1] https://www.watercache.com/education/faq-list

[2] https://www.albanycountyfasteners.com/blog/stainless-steel-and-aluminum/

[3] Andrew Millison https://www.youtube.com/watch?v=DhEaKdmHeCk

[4] Avis and Avis

[5] https://www.rainbrothers.com/store/Downspout-First-Flush-Water-Diverter-Kit-3-p281493345

[6] https://www.harvesth2o.com/filtration_purification.shtml

[7]https://drinking-water.extension.org/drinking-water-treatment-ozone/#How_ozone_treatment_works

Chapter 6

[1]https://www.harvesth2o.com/Dont_Forget_the_Pipe.shtml

[2]https://rainwatermanagement.com/blogs/news/rainwater-harvesting-101-tank-overflow

[3]https://the-chicken-chick.com/the-advantages-of-poultry-nipples/#:~:text=Basically%2C%20as%20long%20as%20you,hens%2C%20again%20as%20a%20minimum.

Chapter 7

[1]https://davenaves.com/blog/how-to/dont-leave-pvc-in-sunlight/

[2]https://www.sahealth.sa.gov.au/wps/wcm/connect/public+content/sa+health+internet/public+health/water+quality/rainwater/rainwater

[3]https://shadesofgreenpermaculture.com/blog/techniques/how-to-winterize-your-water-system/?fbclid=IwAR0vQ7zUUPmZs3ObTuwKnYUvsA93CfGPmqGem005QTw0gowjbKNQqc6pj7Y

ACKNOWLEDGEMENTS

I set out to write this book on my own, thinking it would be a solo project. But it turns out that most projects that you end up being most proud of are only accomplished with the help of a loving team.

To Lorelei Odom and her family: Pure gratitude for your insight into pumps, valves, pipes, and getting your hands messy. I'm so amazed at the progress! I can't wait to check it out myself.

To Lyle Collins: Thank you so much for walking me through the pump math and how to read a pump plot. How little I knew, yet how graciously you explained it all to me! I intend to keep passing the favor forward.

To Delara, Sally, Shannon, Connie, Jill, Shelley, Caydance, Fiona, Sue, Özge, Kelsie, and Kimberley: You ladies are the most amazing group of women I know. Thank you for every morsel of support you all have provided to me—without you all, I probably would have given up. Thank you!

To Lisa Howard, my awesome editor: Thank you for not just editing my book and asking me the right questions, but for mentoring me through the process to find the light again.

To Lulu, my amazing illustrator: For your talents and professionalism, I thank you.

To my family, both new and old: Thank you for supporting me through the dark times and the good times. Thank you for getting excited for me. Thank you for eating my pumpkin pie.

And finally, to my husband: Without you, none of the above could have been accomplished. I say it lot, but here it goes again: I am the luckiest girl in the world.

BIBLIOGRAPHY

[1] D. Pushard, "HarvestH2o.com Incentives," [Online]. Available: https://www.harvesth2o.com/incentives.shtml#pa. [Accessed 23 1 2022].

[2] G. o. S. A. SA Health, "Rainwater | SA Health," 2021. [Online]. Available: https://www.sahealth.sa.gov.au/wps/wcm/connect/public+content/sa+health+internet/public+health/water+quality/rainwater/rainwater. [Accessed 23 1 2022].

[3] U. S. E. P. Agency, "Acid Rain Students Site: PH Scale," Epa.gov, [Online]. Available: https://www3.epa.gov/acidrain/education/site_students/phscale.html#:~:text=Typical%20acid%20rain%20has%20a,acidity%20is%2010%20times%20greater. [Accessed 23 1 2022].

[4] H. C. Alliance, "Water Quality and its Threats | Welcome to Hill Country Alliance," 2018. [Online]. Available: https://hillcountryalliance.org/our-work/water-resources/water-quality-and-its-threats/. [Accessed 23 1 2022].

[5] R. C. D. o. S. C. County, "Stormwater and Erosion Management | RCD Programs," 2022. [Online]. Available: rcdsantacruz.org/rainwater-harvesting-indoor-uses/. [Accessed 23 1 2022].

[6] Home Depot, [Online]. Available: https://www.homedepot.com/p/DRYLOK-Original-1-gal-White-Flat-Latex-Interior-Exterior-Basement-and-Masonry-Waterproofer-27513/100118662?source=shoppingads&locale=en-US&&mtc=Shopping-B-F_D24-G-D24-024_012_WATERPROOFER-Multi-NA-Feed-SMART-NA-NA-FY21_Exterior. [Accessed 23 1 2022].

[7] R. A. a. M. Avis, Essential Rainwater Harvesting: A Guide to Home-Scale System Design, Gabriola Island, BC, Canada: New Society Publishers, 2019.

[8] U. M. G. P. |. U. o. C. A. a. N. Resources, "How Much Water Does My Food Garden Need?," 3 2014. [Online]. Available: https://sonomamg.ucanr.edu/files/185639.pdf. [Accessed 23 1 2022].

[9] Innovative Water Solutions LLC, "The FAQ List," [Online]. Available: https://www.watercache.com/education/faq-list. [Accessed 23 1 2022].

[10] P. Ahrens, "Stainless Steel & Aluminum: Why You Shouldn't Use Them Together and Proper Precautions To Take If You Do - Albany County Fasteners," 15 9 2017. [Online]. Available: https://www.albanycountyfasteners.com/blog/stainless-steel-and-aluminum/. [Accessed 23 1 2022].

[11] A. Millison, "How to HARVEST RAINWATER from your roof," Youtube, 21 9 2021. [Online]. Available: https://www.youtube.com/watch?v=DhEaKdmHeCk. [Accessed 23 1 2022].

[12] Rain Brothers LLC, "Downspout First Flush Water Diverter Kit 3"," 2022. [Online]. Available: https://www.rainbrothers.com/store/Downspout-First-Flush-Water-Diverter-Kit-3-p281493345. [Accessed 23 1 2022].

[13] D. Pushard, "Rainwater - Purification and Filtration," Harvesth2o.com, 2020. [Online]. Available: https://www.harvesth2o.com/filtration_purification.shtml. [Accessed 23 1 2022].

[14] U. C. Extension, "Drinking Water Treatment – Ozone – Drinking Water and Human Health," U.S. Department of Agriculture, 23 8 2019. [Online]. Available: https://drinking-water.extension.org/drinking-water-treatment-ozone/#How_ozone_treatment_works. [Accessed 23 1 2022].

[15] D. Pushard, "Don't Forget the Pipe," Harvesth2o.com, 2022. [Online]. Available: https://www.harvesth2o.com/Dont_Forget_the_Pipe.shtml. [Accessed 23 1 2022].

[16] Rainwater Management Solutions, "Rainwater Harvesting 101: Tank Overflow," 30 4 2018. [Online]. Available: https://rainwatermanagement.com/blogs/news/rainwater-harvesting-101-tank-overflow. [Accessed 23 1 2022].

[17] K. S. Mormino, "Clean Water: The Advantages of Poultry Nipple Waterers | The Chicken Chick®," The Chicken Chick®, 24 7 2012. [Online]. Available: https://the-chicken-chick.com/the-advantages-of-poultry-nipples/#:~:text=Basically%2C%20as%20long%20as%20you,hens%2C%20again%20as%20a%20minimum.. [Accessed 23 1 2022].

[18] D. Naves, "Don't Leave PVC in Sunlight," DaveNaves.com, 24 9 2013. [Online]. Available: https://davenaves.com/blog/how-to/dont-leave-pvc-in-sunlight/. [Accessed 23 1 2022].

[19] Shades of Green Permaculture, "How to Winterize Your Water System," Shades of Green Blog, 29 11 2021. [Online]. Available: https://shadesofgreenpermaculture.com/blog/techniques/how-to-winterize-your-water-system/?fbclid=IwAR0vQ7zUUPmZs3ObTuwKnYUvsA93CfGPmqGem005QTw0gowjbKNQqc6pj7Y. [Accessed 23 1 2022].

PHOTOGRAPHY AND ILLUSTRATION CREDITS

Front Cover

[1] Janice Adlam/Shutterstock.com

[2] © Oleksandr – stock.adobe.com

[3] illustrissima/Shutterstock.com

[4] Gorloff-KV/Shutterstock.com

[5] Animaflora PicStock/Shutterstock.com

Chapter 1

[6] Fellswaymedia/Shutterstock.com

[7] Energy.gov

Chapter 2

[8] dimjing on Fiverr

[9] dimjing on Fiverr

Chapter 3

[10] folihu/Shutterstock.com

[11] K Quinn Ferris/Shutterstock.com

[12] Ronnachai Palas/Shutterstock.com

[13] 135pixels/Shutterstock.com

[14] Dmitry Galaganov/Shutterstock.com

Chapter 4

[15] Tattoboo/Shutterstock.com

[16] Ratchat/Shutterstock.com

[17] Ratchat/Shutterstock.com

[18] pixinoo/Shutterstock.com

[19] pixinoo/Shutterstock.com

[20] stockphotofan1/Shutterstock.com

[21] Janice Adlam/Shutterstock.com

[22] Animaflora PicsStock/Shutterstock.com

[23] Darryl Brooks/Shutterstock.com

[24] Weather.gov

[25] Earth.google.com

[26] pixinoo/Shutterstock.com

Chapter 5

[27] Suzanne Tucker/Shutterstock.com

[28] Douglas Cliff/Shutterstock.com

[29] stockphotofan1/Shutterstock.com

[30] Richard Pratt/Shutterstock.com

[31] © Packshot – stock.adobe.com

[32] Geoff Sperring/Shutterstock.com

[33] Daniel Toh/Shutterstock.com

[34] Douglas Cliff/Shutterstock.com

[35] GSPhotography/Shutterstock.com

[36] tawanroong/Shutterstock.com

[37] Lea Rae/Shutterstock.com

[39] IDostal/Shutterstock.com

[40] Douglas Cliff/Shutterstock.com

[41] zstock/Shutterstock.com

[42] Reabetswe C Matjeke/Shutterstock.com

[43] ENeems/Shutterstock.com

[44] LizzavetaS/Shutterstock.com

[45] Douglas Cliff/Shutterstock.com

[46] r_silver/Shutterstock.com

[47] washarapong hongsala/Shutterstock.com

[48] ivan_kislitsin/Shutterstock.com

[49] niteenrk/Shutterstock.com

Chapter 6

All line drawings: dimjing on Fiverr

[50] Vector illustration: Andrey Apoev/Shutterstock.com

Chapter 7

[51] Douglas Cliff/Shutterstock.com

[52] Valerie Johnson/Shutterstock.com

[53] GSPhotography/Shutterstock.com

[54] Peter Turner Photography/Shutterstock.com

Made in the USA
Las Vegas, NV
15 April 2024